The NASA STI Program Office ... in Profile

Since its founding, NASA has been dedicated to the advancement of aeronautics and space science. The NASA Scientific and Technical Information (STI) Program Office plays a key part in helping NASA maintain this important role.

The NASA STI Program Office is operated by Langley Research Center, the lead center for NASA's scientific and technical information. The NASA STI Program Office provides access to the NASA STI Database, the largest collection of aeronautical and space science STI in the world. The Program Office is also NASA's institutional mechanism for disseminating the results of its research and development activities. These results are published by NASA in the NASA STI Report Series, which includes the following report types:

- TECHNICAL PUBLICATION. Reports of completed research or a major significant phase of research that present the results of NASA programs and include extensive data or theoretical analysis. Includes compilations of significant scientific and technical data and information deemed to be of continuing reference value. NASA's counterpart of peer-reviewed formal professional papers but has less stringent limitations on manuscript length and extent of graphic presentations.

- TECHNICAL MEMORANDUM. Scientific and technical findings that are preliminary or of specialized interest, e.g., quick release reports, working papers, and bibliographies that contain minimal annotation. Does not contain extensive analysis.

- CONTRACTOR REPORT. Scientific and technical findings by NASA-sponsored contractors and grantees.

- CONFERENCE PUBLICATION. Collected papers from scientific and technical conferences, symposia, seminars, or other meetings sponsored or cosponsored by NASA.

- SPECIAL PUBLICATION. Scientific, technical, or historical information from NASA programs, projects, and mission, often concerned with subjects having substantial public interest.

- TECHNICAL TRANSLATION. English-language translations of foreign scientific and technical material pertinent to NASA's mission.

Specialized services that complement the STI Program Office's diverse offerings include creating custom thesauri, building customized databases, organizing and publishing research results ... even providing videos.

For more information about the NASA STI Program Office, see the following:

- Access the NASA STI Program Home Page at *http://www.sti.nasa.gov*

- E-mail your question via the Internet to help@sti.nasa.gov

- Fax your question to the NASA STI Help Desk at (301) 621-0134

- Telephone the NASA STI Help Desk at (301) 621-0390

- Write to:
  NASA STI Help Desk
  NASA Center for AeroSpace Information
  7121 Standard Drive
  Hanover, MD 21076-1320

NASA/TM—2004-211529

EMERGING COMMUNICATION TECHNOLOGIES (ECT) PHASE 3 FINAL REPORT

Gary L. Bastin, Ph.D.
ASRC Aerospace Corporation, John F. Kennedy Space Center, Florida

William G. Harris, PE
ASRC Aerospace Corporation, John F. Kennedy Space Center, Florida

Lakesha D. Bates, B.S.E.E., M.S.E.E.
NASA Graduate Fellowship Student, John F. Kennedy Space Center, Florida

Richard A. Nelson
NASA, YA-D7, John F. Kennedy Space Center, Florida

**National Aeronautics and
Space Administration**

John F. Kennedy Space Center, Florida 32899-0001

September 2004

Acknowledgments

Although there is always the risk of inadvertently forgetting someone, the ECT team nonetheless wishes to acknowledge especially the assistance and guidance provided by the following individuals, listed alphabetically. Without the continued support of these supporters who believed in the value of this project, this project could not have accomplished all its goals.

Name	Organization
Erik Denson	NASA-KSC
Temel Erdogan	ASRC Aerospace Corporation
Richard A. Nelson	NASA-KSC
Don Philp	ASRC Aerospace Corporation
Steve Schindler	NASA-KSC
Lisa Valencia	NASA-KSC

Available from:

NASA Center for AeroSpace Information
7121 Standard Drive
Hanover, MD 21076-1320

National Technical Information Service
5285 Port Royal Road
Springfield, VA 22161

TABLE OF CONTENTS

No.	Description	Page
1.0	**Introduction**	1
1.1	Background	1
1.2	Objectives	2
1.3	Scope	2
1.4	ARTWG / ASTWG	2
2.0	**Free Space Optics**	5
2.1	Background	5
2.2	Basic FSO Theory	7
2.3	Test Description	7
2.4	Test Objectives	8
2.5	Test Setup	8
2.5.1	EDL ANDL Setup	10
2.5.2	EDL East Parking Lot Test Setup	12
2.5.3	EDL to SSPF Parking Lots Test Setup	15
2.5.4	Schwartz Road Test Setup	17
2.6	Test Equipment And Software	19
2.6.1	SONAbeam 622-M	20
2.6.2	Laptop	23
2.6.3	SmartBits	24
2.6.4	SONAbeam Terminal Controller Software	26
2.6.5	SmartApplications	29
2.7	Test Results	30
2.7.1	Results Summary	30
2.7.2	SmartBits	42
2.7.3	SmartBits Testing	42
2.8	FSO Security Concerns	43
2.9	Comparison Of Multi-Beam and Single Beam Systems	44
2.10	FSO Summary And Recommendations	45
3.0	**Ultra Wide Band**	47
3.1	Introduction	47
3.2	Background	49
3.3	OFDM Overview	49
3.4	Research Organization	51
3.5	Timing Jitter in UWB-OFDM Communication Systems	51
3.6	OFDM Technique	51
3.7	Timing Error Effects on System Performance	53
3.8	Timing Jitter and Phase Noise Relationship	53
3.9	Tikhonov Approximation of Timing Error	54
3.10	Bit Syncronization	55
3.11	Analytical Solutions for BER Performance	56
3.12	Conditional Error Probabilities	56
3.13	Manchester Coded Data	57

3.14	NRZ Coded Data	57
3.15	RZ Coded Data	57
3.16	Miller Coded Data	58
3.17	Establishing Avg. Error Prob. at the Receiver	59
3.18	Test Case with the PulsON200 Radios	61
3.19	Test Procedures	62
3.20	Results Overview	65
3.21	BER Effects due to Timing Errors	65
3.22	PulsON200 Test Case	70
3.23	Required Equipment and Test Setup	71
3.24	Test Case Results	71
3.25	Assessment of Timing Uncertainties	73
3.26	Analysis of Effects on UWB-OFDM System with High Data Rates	75
3.27	Summary of Results	76
3.28	Recommendations for Future Research	76
4.0	**ECT Summary Recommendations For Continued Research**	**78**

APPENDIX A: UWB EVK Performance Analysis Tool (PAT) Software 80

APPENDIX B: Digital Signaling Formats Applicable T0 FSO And UWB Communication Signaling 90

APPENDIX C: Q-Function, ERF, AND ERFC, Applicable To Bit Error Rate Theoretical Analysis Of FSO AND UWB Communication Systems 92

REFERENCES 94

LIST OF FIGURES

No.	Description	Page
Figure 1-1	ARTWG/ASTWG National Development Strategy	3
Figure 1-2	ARTWG Integration/Interaction Process	3
Figure 1-3	Spaceport and Range Environments	4
Figure 2-1	Typical Test Setup Using SmartBits with Loop at OTU #2	9
Figure 2-2	OTU #1 in the ANDL on the North Bench	10
Figure 2-3	OTU #2 in the ANDL by the South Wall	11
Figure 2-4	OTU #1 with Laptop Measuring Receiver Performance	12
Figure 2-5	OTU #1 Mounted on an Existing Antenna Trailer	13
Figure 2-6	OTU Mounted On Fabricated Support Stand	13
Figure 2-7	Trailer Mounted OTUs in East EDL Parking Lot (113-Ft Range)	14
Figure 2-8	Portable Generator Used for Power During Remote Testing	14
Figure 2-9	EDL to SSPF Parking Lots Test Setup	15
Figure 2-10	OTU #2 in the EDL North Parking Lot	15
Figure 2-11	OTU #1 in the SSPF Parking Lot	16
Figure 2-12	View from OTU #2 to OTU #1	16
Figure 2-13	OTU #2 Parked on West End of Schwartz Rd.	17
Figure 2-14	OTU #1 Parked on East End of Schwartz Rd.	18
Figure 2-15	OTU #1 with SmartBits and Laptop Connected	18
Figure 2-16	View from OTU #1 toward OTU #2 at 1.0 Mile Range	19
Figure 2-17	Front of SONAbeam 622-M OTU	20
Figure 2-18	SONAbeam Junction Box Open	22
Figure 2-19	Pole Mount with OTU, Connection Box & Power Supply Box	22
Figure 2-20	Gateway 450 XL Laptop Computer	23
Figure 2-21	SmartBits Test Unit	25
Figure 2-22	Field Testing with the SmartBits & a Laptop	25
Figure 2-23	Initialization Page for fSONA Terminal Controller Software	26
Figure 2-24	Comm 1 Status Showing No FSO Connection or Input	26
Figure 2-25	Comm 1 Tx Configuration With System Power At Level 1	27
Figure 2-26	Comm 1 Rx Power Level	27
Figure 2-27	Comm 1 Diagnostics with No FSO Lock or Input	28
Figure 2-28	Comm 1 Settings	28
Figure 2-29	Smart Applications Main Page with Card 17 to Card 19 Test Setup	29
Figure 2-30	Minimum Receive Power versus Distance	32
Figure 2-31	Receive Power versus Transmitter Setting @ 1.0 Mile	32
Figure 2-32	Receive Power versus Transmitter Setting @ 1.5 Mile	33
Figure 3-1	De-Multiplexed High Data Rate Of DM Data Stream	49
Figure 3-2	Overlapping Orthogonal Sub-Carriers In OFDM Symbol	50
Figure 3-3	OFDM Symbol	52
Figure 3-4	OFDM Block Diagram	54
Figure 3-5	Performance Analysis Interface For PulsON 200 Receiver	63
Figure 3-6	Theoretical BER vs Eb/No For PulsON 200 Flip Modulation	64

Figure 3-7	Representation Of Timing Errors In A Digital Signal	66
Figure 3-8	Representation Of Distortions In A Signal	66
Figure 3-9	BERs For Timing Jitter (Manchester Data)	67
Figure 3-11	BERs For Timing Jitter (MILLER Data)	68
Figure 3-12	BERs For Timing Jitter (NRZ Data)	69
Figure 3-13	Flip Modulation	70
Figure 3-14	PulsON 200 Evaluation Kit Setup	71
Figure 3-15	Flip Modulation Theoretical Curve vs PulsON 200 Data (9.6 Mbps)	73
Figure 3-16	Manchester Data With Extrapolated PulsON 200 Data	74

LIST OF TABLES

No.	Description	Page
Table 2-1	FSO Industry Comparisons	6
Table 2-2	FSO Test Locations	9
Table 2-3	SONAbeam 622-M Specifications	21
Table 2-4	FSO Factory IP Addresses	23
Table 2-5	FSO Reset IP Addresses	23
Table 2-6	Laptop Computer Specifications	24
Table 2-7	Laptop Computer Installation Parameters	24
Table 2-8	Summary of Test Locations & Objectives	30
Table 2-9	Summary of Test Types and Data Measurements	30
Table 2-10	Recommended Receiver Power Levels	31
Table 2-11	Summary Of fSONA 622-M Testing	33
Table 2-12	SmartBits Throughput Test Parameters	42
Table 2-13	Typical SmartBits Through-Put Test	43
Table 2-14	Typical SmartBits Packet Loss Test	43
Table 3-1	List Of UWB-Related Abbreviations	48
Table 3-2	Analytical Equation Definitions	60
Table 3-3	PulsON 200 Collection Of Data Rate Vs Range Test Cases	72

Executive Summary

The National Aeronautics and Space Administration (NASA) is investigating alternative approaches, technologies, and communication network architectures to facilitate building the Spaceports and Ranges of the future. These investigations support the Crew Exploration Vehicle (CEV) and other associated craft presently under development or under consideration in Government, academic, and private sectors. These investigations also provide a national centralized R&D forum for next-generation Spaceport and Range technology development. Together, these sectors all share the common goal of changing the historic risk/reward equation for access to space, with the intent to:

- Dramatically reduce launch cost
- Greatly improve launch system reliability
- Significantly reduce crew risk

ECT - Phase 3

1.0 <u>INTRODUCTION</u>

1.1 **BACKGROUND**

Emerging Communication Technologies is a multi-year task investigating new communication technologies with likely high utility and application for future ranges and spaceports.

In year one, the project was called Range Information Systems Management (RISM). This project investigated US ranges and documented their missions, capabilities, and infrastructures. A part of this investigation was the history of communication and the identification of certain technologies which might offer improved range capabilities in the near future. Three emerging technologies were identified: Free Space Optics (FSO), Ultra Wide Band (UWB), and Wireless Ethernet (Wi-Fi). All three of these technologies address the first mile / last mile communication solution.

In year two, specific examples of FSO, UWB and Wi-Fi were purchased and evaluated for range application. The FSO hardware was an AirFiber 5800 optical transceiver. This unit includes an auto-tracking feature that keeps the two units in optical alignment during small movements that normally occur to support structures due to solar heating, winds and vibration. The units were tested over various distances and through various weather conditions.

The UWB effort in year 2 involved an industry survey, a discussion of UWB theory, and the testing of an Evaluation Kit (EVK) from Time-Domain Corporation. Testing involved measuring the signal degradation due to range; normal office barriers of concrete, metal, partitions, etc.; and interference from microwave ovens, wireless phone, etc.

During year 2, two different Wi-Fi systems were purchased and evaluated. One was an 802.11b base station by Microsoft. The other was an 802.11g system by D-Link. The majority of testing was on the 802.11b system were signal degradation due to range; normal office barriers of concrete, metal, partitions, etc.; and interference from microwave ovens, wireless phone, etc., were investigated.

During year 3, the current year, additional details were addressed in detail and are documented at length in this report. Specifically, wide-beam, non-tracking FSO hardware from fSONA was procured and evaluated in contrast to the narrow-beam, auto-tracking hardware from AirFiber investigated in year 2. Additionally, OFDM-UWB performance limitations were investigated through using the same EVK with the addition of new firmware upgrades from Time-Domain Corporation. Likewise, general industry trends for achieving first mile / last mile communications were monitored and incorporated into this multi-year ECT activity.

1.2 OBJECTIVES

The primary objective for the Emerging Communication Technology (ECT) task is to lead the development of a Space Based Range Distributed Subsystem (SBRDS) network providing the concurrent features and growth capabilities necessary for future Spaceports and Ranges to interconnect Range assets, Range operations, and Range users during launch and recovery events, while focusing primarily on the First Mile/Last Mile wireless communication extensions to existing, fixed communication infrastructures.

1.3 SCOPE

ECT Phase 3 was limited to the following:
- Investigate one fixed alignment, multiple-beam FSO system
- Further investigate the capabilities of the Time-Domain upgraded firmware UWB system
- Monitor future range and spaceport architectures and needs

1.4 ARTWG / ASTWG

ECT further seeks to multiply the knowledge base of the in-house investigators through participation in the active efforts of:
- ARTWG (Advanced Range Technology Working Group)
- ARTWG Communication Subgroup
- ARTWG other Subgroups
- ASTWG (Advanced Spaceport Technology Working Group)
- ASTWG Subgroups

ARTWG is a collaborative NASA/US Air Force/Industry/Academia effort to focus interest and investment in Range technologies (Figure 1-1). It is co-chaired by NASA and the US Air Force, and comprised of aerospace leaders from industry, academia, and national, state, and local governments. ARTWG is a multi-layer organization with functional subgroups as its base (Figure 1-2). ARTWG addresses Range (Figure 1-3) development needs while its companion organization ASTWG (Advanced Spaceport Technology Working Group) addresses Spaceport development needs.

Figure 1-1 ARTWG/ASTWG National Development Strategy

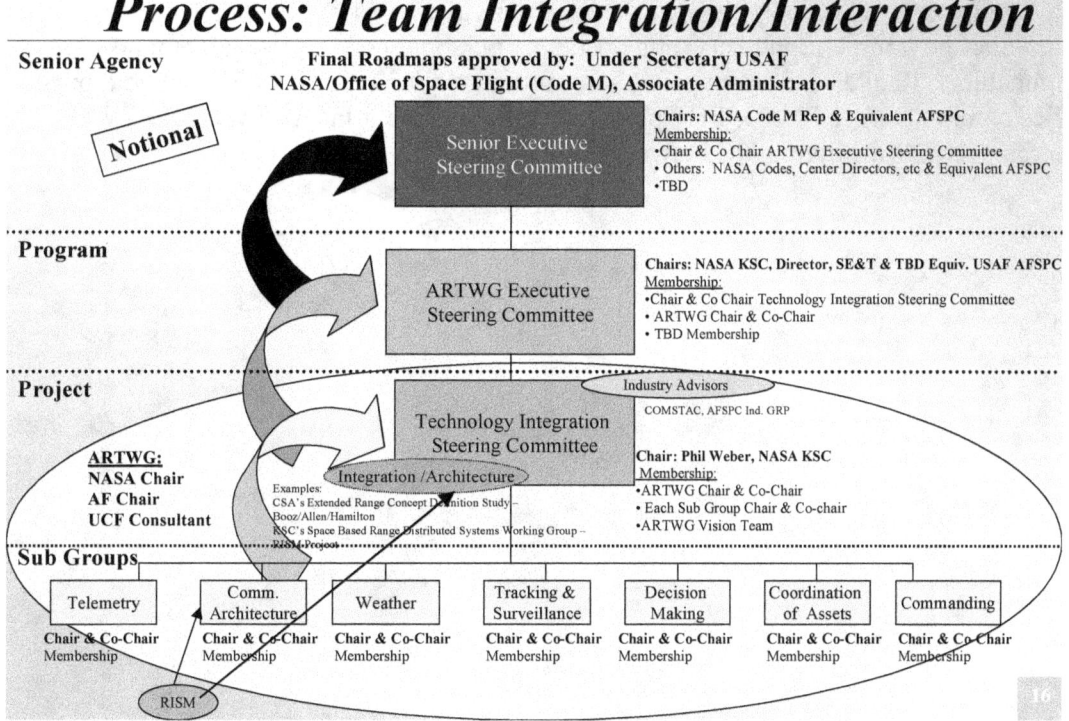

Figure 1-2 ARTWG Integration/Interaction Process

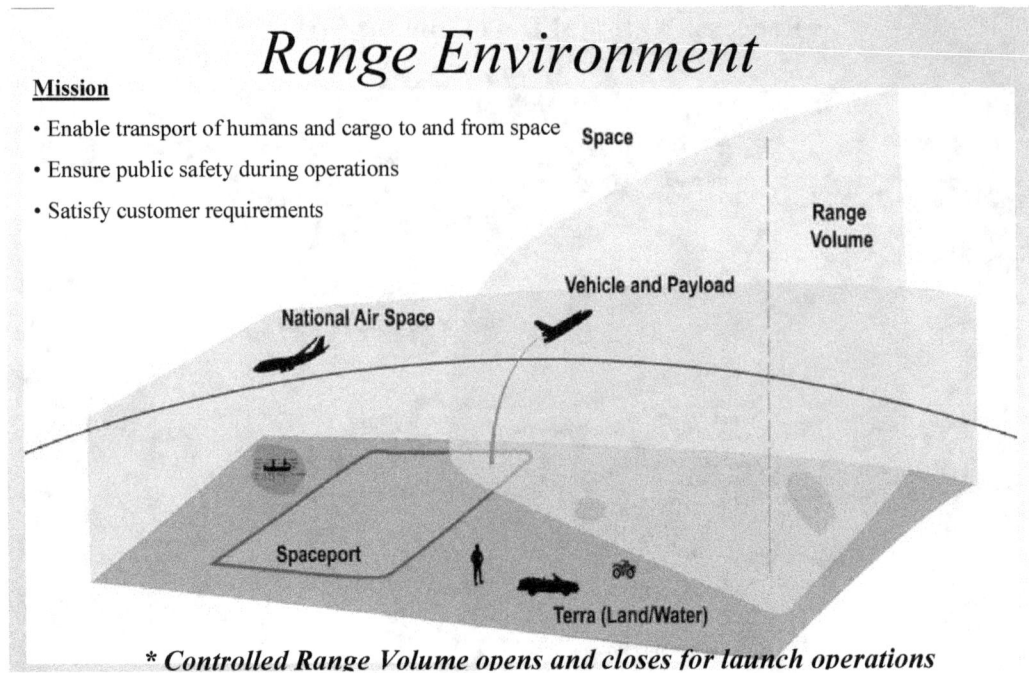

Figure 1-3   Spaceport and Range Environments

ARTWG and ASTWG complement the ECT activities and provide an avenue to interact with other government, industry, and academia personnel on new developments applicable to Ranges and Spaceport of the future. National conferences are held once or twice a year. During FY-03, ECT personnel attended the joint ARTWG / ASTWG conference in Washington DC between May 24 and May 26.

2.0 FREE SPACE OPTICS

2.1 BACKGROUND

Free Space Optics (FSO) was one of the original three First Mile/Last Mile broadband wireless access systems identified in the RISM Phase I report[1]. An auto-track system by AirFiber was purchased, tested and reported during the ECT Phase 2 activities[2]. For ECT Phase 3, a fixed-alignment and multiple beams system by fSONA was procured and tested.

Optical communication systems provide the highest available carrier frequencies and thus the fastest data rates possible today. FSO is designed to be a lower cost alternative to conventional fiber-optic cable-based communication links[3]. FSO is especially attractive within a metropolitan environment where the costs for trenching, cable installation, and street repairs can run from $200K to easily over $1M per mile, depending on the urban location.

FSO is a maturing technology that offers significant enhancements over most wireless technologies, including higher data rate, and the complete avoidance of any spectrum licensure costs. Its primary competition today is from existing fixed fiber installations. Today, a significant percentage of FSO sales are international. This has occurred due to the extensive USA fiber infrastructure that was installed in the 1990's slowing its expansion within the USA.

Although FSO offers the potential of maximum wireless performance, the limited opportunities within the US and an international recession have combined to reshuffle the FSO industry. Table 2-1 lists and compares key international FSO players. AirFiber, the industry leaders 2 years ago but presently out of business[4], is listed for comparison.

FSO links are based on infrared lasers and optical detectors. Over short distances, they are capable of providing very high data rates. Standard rates of OC-3 (155 Mb/s), OC-12 (622 Mb/s), and OC-48 (2.5 Gb/s) are all available off the shelf today.

The primary limitations on using FSO involve weather over distance. Thick fog can attenuate the laser signals and restrict the usefulness of a FSO link. Distance is not normally a concern when FSO is used as a "First Mile" technology; however, distance does magnify weather effects.

---

[1] Range Information Systems Management (RISM) Phase I Report, NASA/TM-2004-211523, September 2002
[2] Emerging Communication Technologies (ECT) Phase 2 Report, Volume 1, Main Report, NASA/TM-2004-211522, September 2003
[3] http://www.airfiber.com/products/index.htm
[4] http://www.airfiber.com is operation and was updated in 2004

Table 2-1   FSO Industry Comparisons

Company	Tx	Auto Track	Wave Length	Comments
AirFiber	1	Y	785	Out of business
Alcatel SA	-	-	-	Uses fSONA equipment
Cablefree Solutions	3	N	780	UK
Canon Inc.	1	Y	785	USA
Communication By Light GmbH (CBL)	4	N	870	Germany
Corning Cable Sys	4	Y	850	USA
Dominion Lasercom	1	Y	850	USA
fSona Com	4	N	1550	Canada
Infrared Com Systems (ICS)	1	N	780 & 980	USA
Infrared technologies America, LLC	-	-	-	Uses LaserBit equipment
iRLan Ltd.	1	N	unk	Israel
LaserBit Com	8	N	unk	USA.
LightPointe Communications	4	Y	850	USA.
LSA Photonics	1	N	785	USA
Maxima Corp	unk	unk	10,000	Long wave length infrared; USA
Mostcom Ltd	-	-	-	Same as Sceptre
MRV (TeraScope)	4	N	850	USA
Omnilux Inc.	3	unk	unk	Mesh design, USA
PAV Data Systems	3	N	830	UK
Plaintree Systems	4	N	unk	Canada
Sceptre Comm. Ltd.	2	N	850	UK & Russia
Terabeam Corp.	1	Y	1550	USA

FSO testing under the ECT project was conducted around the fSONA SONABEAM 622-M transceiver units. The fSONA system was selected after an evaluation of multi-beam FSO COTS equipment and manufacturers. The fSONA system was selected based on the following:

- Cost
- Engineering & Factory Support
- Market Share
- Technical and Performance Features
- Training

2.2   BASIC FSO THEORY

Free Space Optical (FSO) communication was discussed at length in the previous Phase I RISM report and the ECT Phase 2 report. A brief summary is repeated here for continuity, and to serve as an introduction to the technology for those unfamiliar with the technology.

FSO dates to pre-history. Extensive FSO networks were established in the 19th Century throughout France and North Africa, based around semaphore systems. Later, during the latter part of the 19th century, FSO telephone communication was developed.

The modern FSO age commenced with the invention of the laser slightly more than 40 years ago. Coherent light provided the ability to select specific wavelengths to achieve FSO systems that enable lessening atmospheric attenuation, providing operation through rain, and achieving eye-safety through the selection of appropriate wavelengths for the laser light selected.

Fundamentally, modern FSO systems typically employ NRZ (non-Return to Zero) modulation of laser light. Digital data is encoded as either a high intensity beam or as a low intensity beam, depending on the extinction ratio present in the ON to OFF states engendered by the modulating device.

Within the receiver, a photodetector provides the optical to electrical (OE) conversion. Depending on the range over which communication is desired, both Positive-Intrinsic-Negative (PIN) diodes and APDs (Avalanche Photodetector Diodes) are used. PIN diodes provide less sensitivity, but require only minimal voltage bias to make them operational. APDs provide the maximum in sensitivity, but require voltages often exceeding 100 Volts dc to achieve their maximum sensitivity. This, in turn, increases the need for properly coating circuit cards for FSO apparatus intended for use outdoors, through conformal coating the cards, in order to avoid accidentally shorting out the high dc bias during high humidity conditions.

At the output of the photodetector is a Trans-Impedance Amplifier (TIA). Its purpose is to provide the necessary gain by which to generate a voltage from the current produced by the photodetector diode when exposed to light. Beyond this lie the framing and other packetizing electronics, needed to provide the proper data interfaces for the subsequent parts of the communication system. For fiber optic extensions, it is necessary to have clock and data recovery circuitry, by which clocks are derived from incident light pulses coming into the FSO system via fiber optic cable, to provide proper timing interfaces.

For the fSONA system tested on this project, a Smartbits OC-12 fiber optic interface operating at 622 Mb/s served as the physical data interface in and out of the two units.

2.3   TEST DESCRIPTION

Phase 3 FSO testing was to evaluate a fixed alignment, multi-beam FSO system. The SONAbeam 622-M system was selected. Initial tests were to become familiar with the system and to evaluate its performance at short distances. Later tests were to evaluate performance at increasing link distances. Weather testing was not accomplished due to the combination of lack of adequate weather condition opportunities, safety concerns of lightning-free rainy weather, and storm closures during Hurricanes Charlie and Frances which closed KSC for a total of 7 days during the last two months of the project during multiple weeks.

2.4 TEST OBJECTIVES

The FSO test objectives were as follows:
- Evaluate COTS FSO equipment for possible future use at KSC
- Identify any fundamental shortcomings that must be filled in commercial FSO communication technologies prior to integrating this technology into future range architectures.
- Evaluate a fixed alignment / multiple beam system against the previously tested auto-tracking single beam system.

2.5 TEST SETUP

Various test setups were utilized in evaluating the FSO equipment. The basic setup was to place the two Optical Transfer Units (OTUs) a fixed distance apart and to establish a link. Initial alignment was usually accomplished using the factory-provided 9X power rifle scope with eye-safe internal filtering at the 1310 nm wavelength transmitted by the transmitting lasers.

Once an initial link was established, the link was refined and optimized using the factory provided software on a laptop computer. This provided a display of micro-watts at the OTU receiver. Fine adjustment was accomplished using jacking screws on both azimuth and elevation lever arms. Once the alignment was complete, the azimuth and elevation axis were clamped and not changed until the OTU were repositioned.

A SmartBits Bit Error Rate test device was used to determine throughput and packet loss for various packet sizes. A typical setup is shown in Figure 2-1. A SmartBits sent a variable length data stream to the input of OTU #1. The laser transmitters at OTU #1 transferred the data packets to OTU #2. The output from OTU #2 was connected to its input. OTU #2 then used its laser transmitter to return the data to OTU #1. The output from OTU #1 was returned to the SmartBits where a comparison was made to determine any throughput or data packet losses.

The primary independent variable for all tests was link distance. A summary of test locations and distances is included in Table 2-2. Specific details about each test location are included in the following sections. The FSO units, test equipment and software are described in later sections.

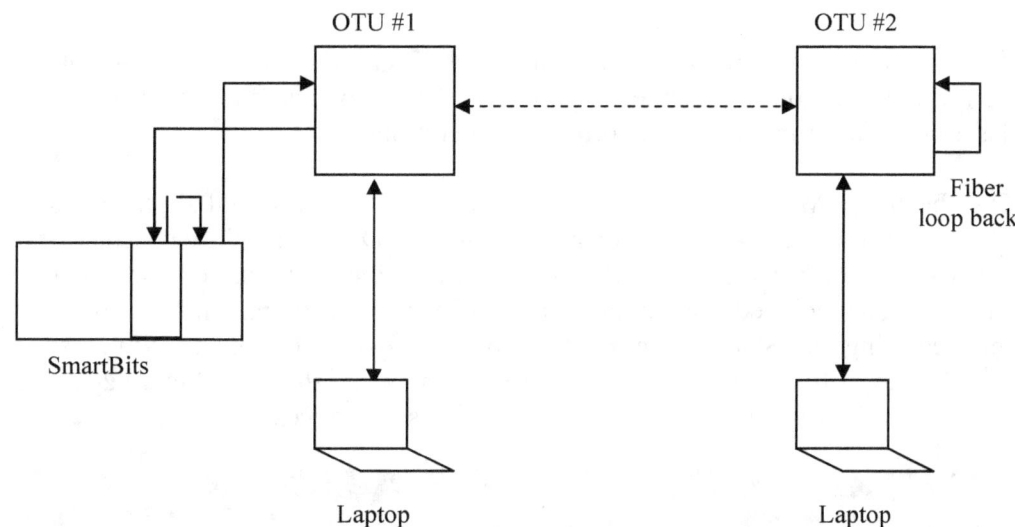

Figure 2-1 **Typical Test Setup Using SmartBits with Loop at OTU #2**

Table 2-2   FSO Test Locations

No.	Location	One Way Distance	Loop Back
1	EDL ANDL[5]	28 ft	Y/N
3	EDL East Parking Lot	113 ft	Y
4	EDL to SSPF Parking Lots	1066 ft	Y
5	Schwartz Road	1.0 mile	Y
6	Schwartz Road	1.5 mile	Y
7	Schwartz Road	1.75 mile	Y
8	Schwartz Road	2.0 mile	Y
9	Schwartz Road	2.5 mile	Y

---

[5] EDL Advanced Network Development Lab, Rm 124

2.5.1 EDL ANDL Setup

Initial testing and checkout were performed in the Advanced Network Development Lab (ANDL), Room 124 in the Engineering Development Lab (EDL). A pair of short pipe stands was fabricated to enable installation of the fSONA 622M units within the ANDL. Universal electrical boxes were modified to house the power supplies which converted 115 VAC to 48 VDC.

OTU #1 was placed on top of an existing workbench (Figure 2-2) while OTU #2 was place 30 feet away on top of a cabinet (Figure 2-3). The two locations were selected to avoid laser beam interruption during normal lab operations.

The SONAbean 622-M is a fixed alignment, multi-beam design. Initial alignment is normally achieved using an eye-safe filtered rifle scope. Due to the close confines of the ANDL, it was impossible to use the rifle scope. Initial alignment was accomplished in the ANDL using a machined adapter and a laser pointer. Fine alignment was still accomplished using the fSONA Terminal Controller Software. For most ANDL tests, only a single transmitter was used. A fiberglass screen mesh, shown in Figure 2-2, was used over the receiver to attenuate the signal at the close distances.

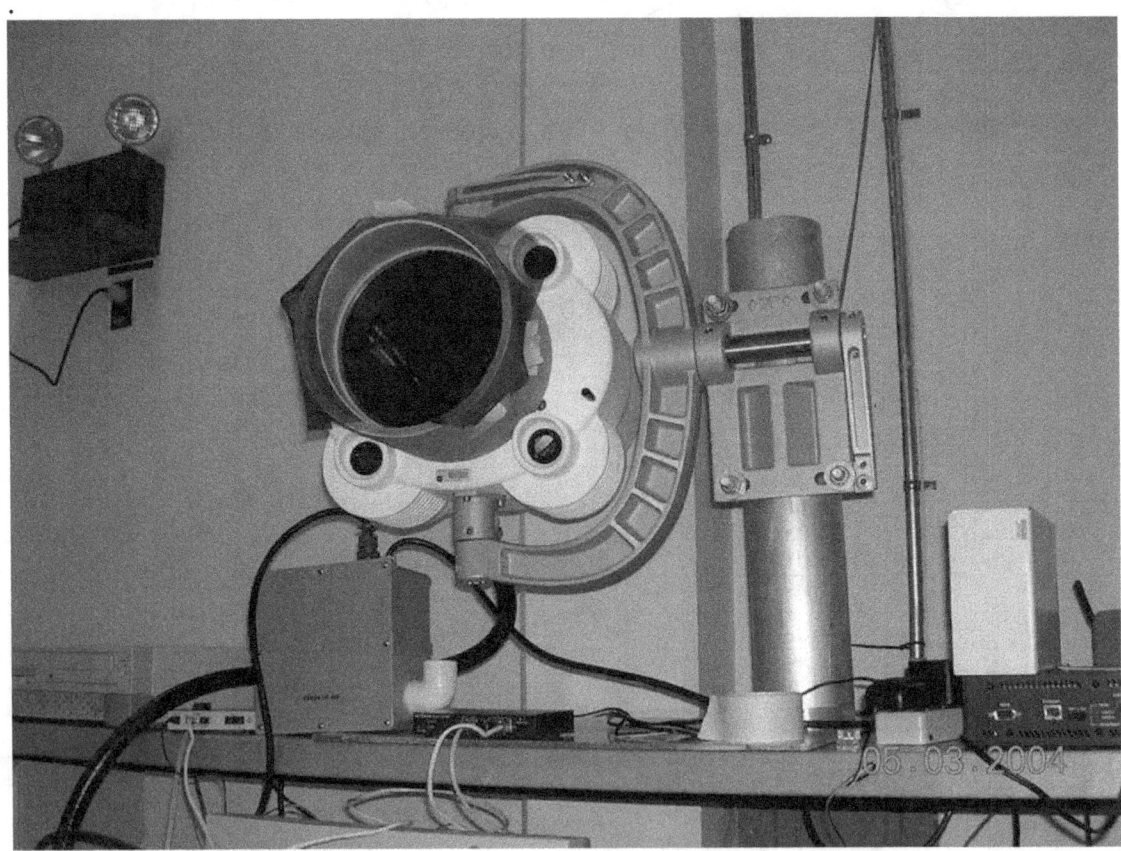

Figure 2-2 OTU #1 in the ANDL on the North Bench

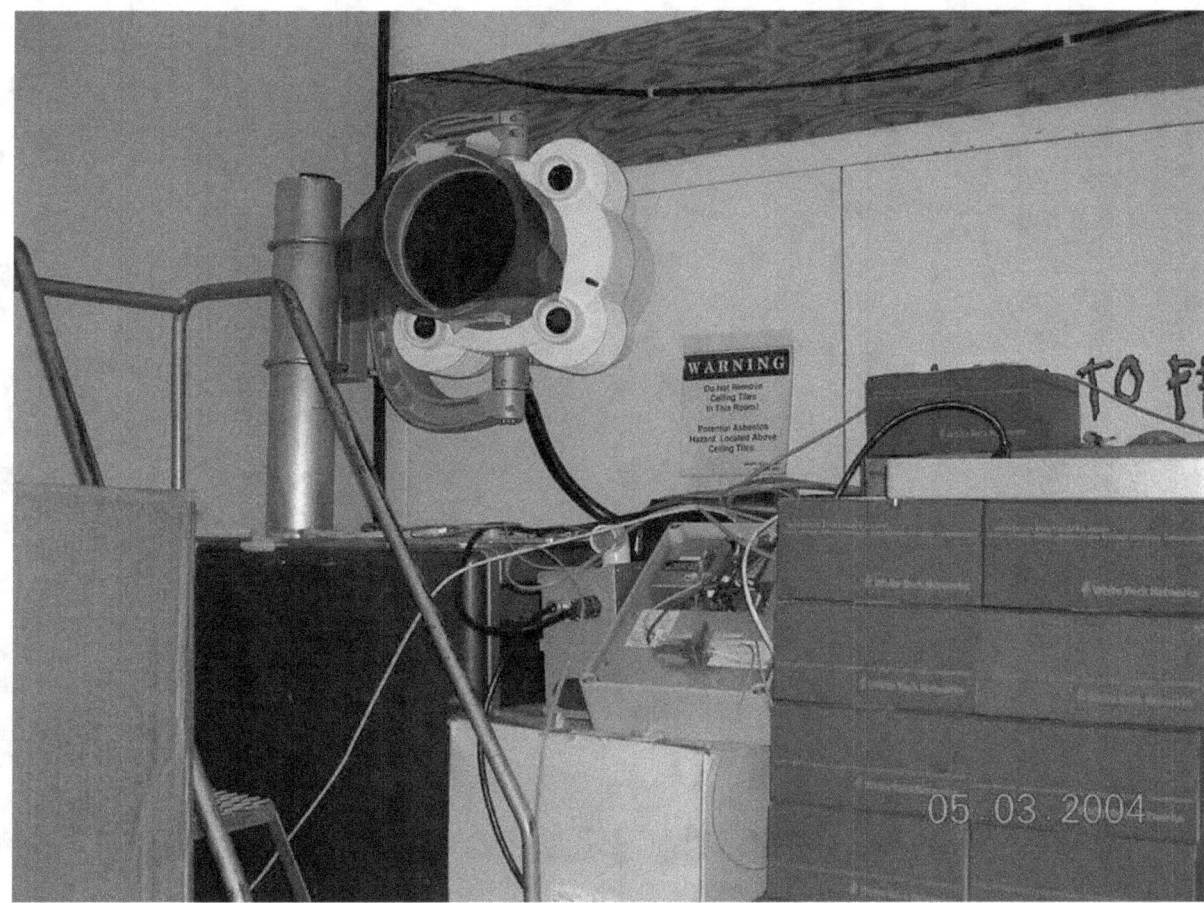

Figure 2-3 OTU #2 in the ANDL by the South Wall

The Advanced Network Development Lab was chosen for initial checkout and testing since it was a secure area and contained a rack-mounted SmartBits test unit that could be used as a source of optical data packets. The SmartBits unit was configured as shown in Figure 2-1 for some tests and configured for a single optical pass in other tests. In these latter tests, the fiber loop-back at OTU #2 was replaced with a direct fiber link back to the SmartBits. Multi-mode fiber (MMF) was used for all cables to and from the SmartBits and for the loop back. A 10 dB optical attenuator was required on the fiber output to OTU #1 in order for the SmartBits to work properly. This remained in place for all tests outside the ANDL.

Testing was accomplished by measurements the receiver power using a laptop running the fSONA Terminal Controller Software. A typical setup is shown in Figure 2-4.

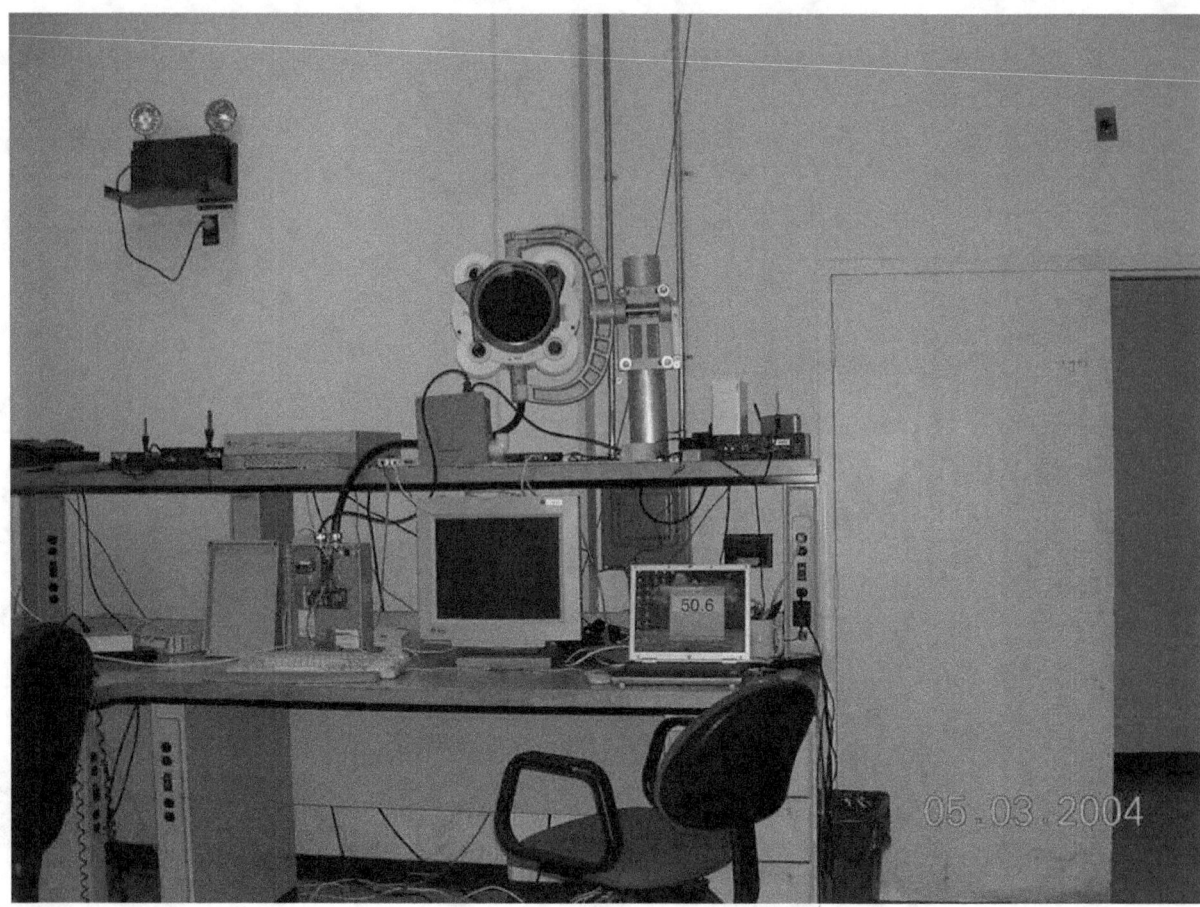

Figure 2-4   OTU #1 with Laptop Measuring Receiver Performance

2.5.2   EDL East Parking Lot Test Setup

The second series of FSO tests were performed on the East parking lot of the EDL. To facilitate moving, spacing and positioning the units, each FSO units was mounted on a trailer. A pair of existing antenna trailers (Figure 2-5), remaining from another project, were modified to support FSO testing. New tri-pod mounts were fabricated and bolted to the trailer structure to provide a quick mount (Figure 2-6). The antennas shown stored horizontally across the top of the trailers were not used in these tests.

Figure 2-5 OTU #1 Mounted on an Existing Antenna Trailer

Figure 2-6 OTU Mounted On Fabricated Support Stand

The testing configuration was per Figure 2-1. OTU #1 and OTU #2 were located 154 feet apart (Figure 2-7). This distance was selected based on available parking space. Power was provided by a pair of small generators (Figure 2-8).

Figure 2-7 Trailer Mounted OTUs in East EDL Parking Lot (113-Ft Range)

Figure 2-8 Portable Generator Used for Power During Remote Testing

2.5.3 EDL to SSPF Parking Lots Test Setup

The third series of FSO tests were between the Engineering Development Laboratory (EDL) and the Space Station Processing Facility (SSPF) parking lots (Figure 2-9). The trailer with OTU #2 was parked on the North side of the EDL (Figure 2-10). The second trailer with OTU #1 was parked in the SE corner of the SSPF East parking lot (Figure 2-11). Figure 2-12 is a view from the OTU #2 toward the OTU #1. The test configuration was again per Figure 2-1. This setup provided a one-way distance of 950 feet with a round-trip distance of 1900 feet.

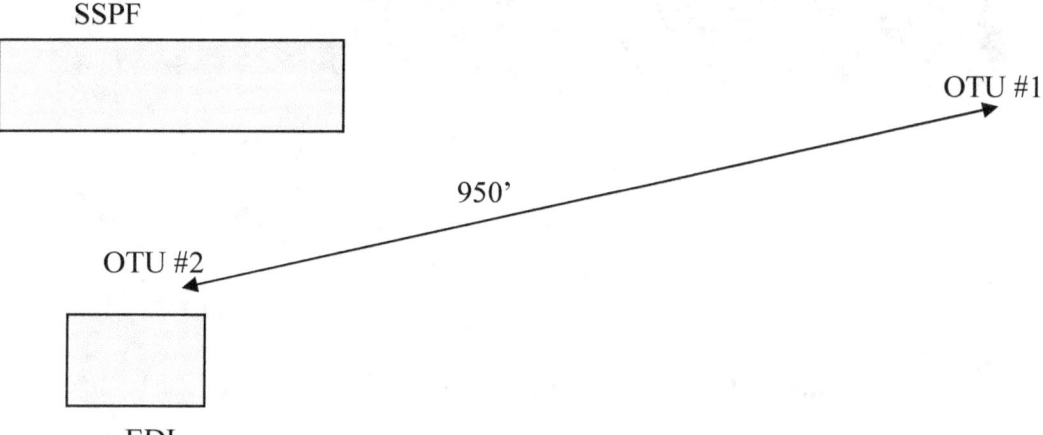

Figure 2-9 EDL to SSPF Parking Lots Test Setup

Figure 2-10   OTU #2 in the EDL North Parking Lot

Figure 2-11   OTU #1 in the SSPF Parking Lot

Figure 2-12   View from OTU #2 to OTU #1

2.5.4 Schwartz Road Test Setup

Schwartz Road is a remote East-West road at KSC that runs relatively straight for 2.7 miles. Extensive long-distance testing was performed at this location. Trailers were positioned off to the side of the road at various distances. Initial testing was performed at 1.0 mile. This was followed with tests at 1.5, 2.0, 2.5, 2.0, 1.75, and 1.5 miles respectively. The West trailer (OTU #2, Figure 2-13) remained relatively fixed for all tests. The East trailer with OTU #1 (Figure 2-14) was repositioned as needed to set the desired link testing distance. The test configuration was again per Figure 2-1 with the SmartBits and Laptop positioned on the trailer (Figure 2-15). The longer distances tested on Schwartz Rd. often made it difficult to even see the mating trailer. Figure 2-16 shows the view from the one mile range. Tests at 2.5 mile were even more demanding during initial alignment.

Figure 2-13   OTU #2 Parked on West End of Schwartz Rd.

Schwartz Road testing was interrupted twice due to hurricanes. Hurricane Charley required that the trailers be moved to nearby fixed objects where they could be tied down. A few weeks later saw the arrival of Hurricane Frances.

Figure 2-14   OTU #1 Parked on East End of Schwartz Rd.

Figure 2-15   OTU #1 with SmartBits and Laptop Connected

Figure 2-16   View from OTU #1 toward OTU #2 at 1.0 Mile Range

2.6   TEST EQUIPMENT AND SOFTWARE

Key FSO test hardware included the following:

- SONAbeam 622-M - Optical Transceiver Unit (OTU)
- Laptop – Laptop computer for interfacing with the OTU
- SmartBits – Data packet source for measuring Throughput and Packet Loss

In addition to the above hardware, two software packages were instrumental in testing and data acquisition. These software packages were:

- SONAbeam Terminal Control Software
- SmartApplications

2.6.1 SONAbeam 622-M

The SONAbeam 622-M OTU shown in Figure 2-17 was the primary component under test. Specifications for the units are summarized in Table 2-3. A pair of 622-M units was purchased around 3/1/04 under the ECT task order (#00087). The purchase price included the following:

- (2) 622-M OTUs
- (2) connection boxes
- (2) Power supplies
- (2) Tri-pods
- Associated equipment
- Terminal controller Software
- Factory training for 2

Figure 2-17   **Front of SONAbeam 622-M OTU**

Table 2-3   SONAbeam 622-M Specifications

Manufacturer		fSONA Communication Corp 140-11120 Horseshoe Way Richmond, B.C. Canada
Model		SONAbeam 622-M
Cost		$40,522.69/pair w/training
Purchase Date		3/15/04
Data rate		OC-12 (622 Mbps)
Distance	Min	30 ft
	Max	1.6 miles; 1.5 miles successfully tested
Transmitters	No.	4
Tx	Wave length	1550 nm
Receiver	Dia	8-inch
Interfaces Types	Fiber	Single mode or Multi-mode
Interfaces	Connector	SC
	Management	RJ-45 or DB9
Voltage		-48 vdc
BER		10^{-12}
Environment	Max Operating Temp	140 F
	Max Operating wind	100 mph
Warranty		3 years
Laser Safety		Class 1M
Serial Numbers	#1	1130050858
	#2	1130030756

Factory training was included with the purchase of the 622-M. Training took place in Vancouver B.C. on 2/18/04. ASRC employees Dr. Gary Bastin and Bill Harris attended.

The units came with a factory-supplied junction box. The inside of the box is shown in Figure 2-18. Payload fiber, management Ethernet, and power connections are made within this box. The two orange fibers running up to the right are the multimode fibers connecting the SmartBits to OTU #1. Also visible on the lower fiber is the 10 dB attenuator inserted in the output line to enable the OTU and SmartBits to communicate. Figure 2-19 shows the junction box mounted on the support pole just below the OTU. The smaller utility box in this figure is the in-house fabricated power supply housing. It contained the 110 VAC to -48 VDC power supply.

Figure 2-18   SONAbeam Junction Box Open

Figure 2-19   Pole Mount with OTU, Connection Box & Power Supply Box

Each OTU was shipped with a factory assigned address (Table 2-4). These were later field changed as shown in Table 2-5.

Table 2-4   FSO Factory IP Addresses

OTU	#1	#2
IP Address	0.0.0.1	0.0.0.1
Subset Mask	255.255.255.0	255.255.255.0
Gateway	192.168.1.254	192.168.1.254

Table 2-5   FSO Reset IP Addresses

OTU	#1	#2
IP Address	128.217.108.178	128.217.108.179
Subset Mask	255.255.255.0	255.255.255.0
Gateway	128.217.108.10	128.217.108.10

2.6.2   Laptop

A pair of Gateway laptop computers, as shown in the following figure, were used to support the ECT testing. Each computer was loaded with fSONA's Terminal Controller Software. This software enabled each OTU to be initialized, controlled, and monitored. The software is discussed in a later section.

Figure 2-20   Gateway 450 XL Laptop Computer

Specifications for the Laptops are shown in the following table.

Table 2-6   Laptop Computer Specifications

Manufacturer	Gateway
Model	DS 450 XL
Processor	Intel Pentium 4
Speed	2.0 GHz
Hard Drive	40 GB
RAM	512 MB
Connectors	USB, RJ-45, Phone
Wi-Fi Standard	802.11b (Internal)
Operating System	Windows XP V.2002

Table 2-7   Laptop Computer Installation Parameters

Name	BH	GB
IP Address	128.217.107.119	128.217.107.175
MAC	00-02-2D-6E-A2-F4	00-02-2D-6E-5B-7E

2.6.3   SmartBits

An existing SmartBits test unit, the lower unit shown in Figure 2-21, was used to test the OTUs. The SmartBits created varying size data packets that were sent through the FSO communication link at OC-12 data rates. The SmartBits compared the data sent with the data received and produced a report on Throughput and Packet Loss.

The SmartBits are populated with test drivers that produce data streams of different protocols. For ECT testing, Cards 17 and 19 shown in Figure 2-21 were used. These cards are for ATM at OC-12. Card 19 was usually the transmitter and Card 17 was the receiver. The jumper in Figure 2-21 was used from the receive port in Card 19 to the transmit port of Card 17.

The SmartBits was initially rack mounted in the ANDL. During lab testing, a duplex multimode fiber was routed from the input and output ports of the OTUs. For testing in the parking lots and on Schwartz Rd., the SmartBits was removed from the rack (Figure 2-22) and powered by the portable generator.

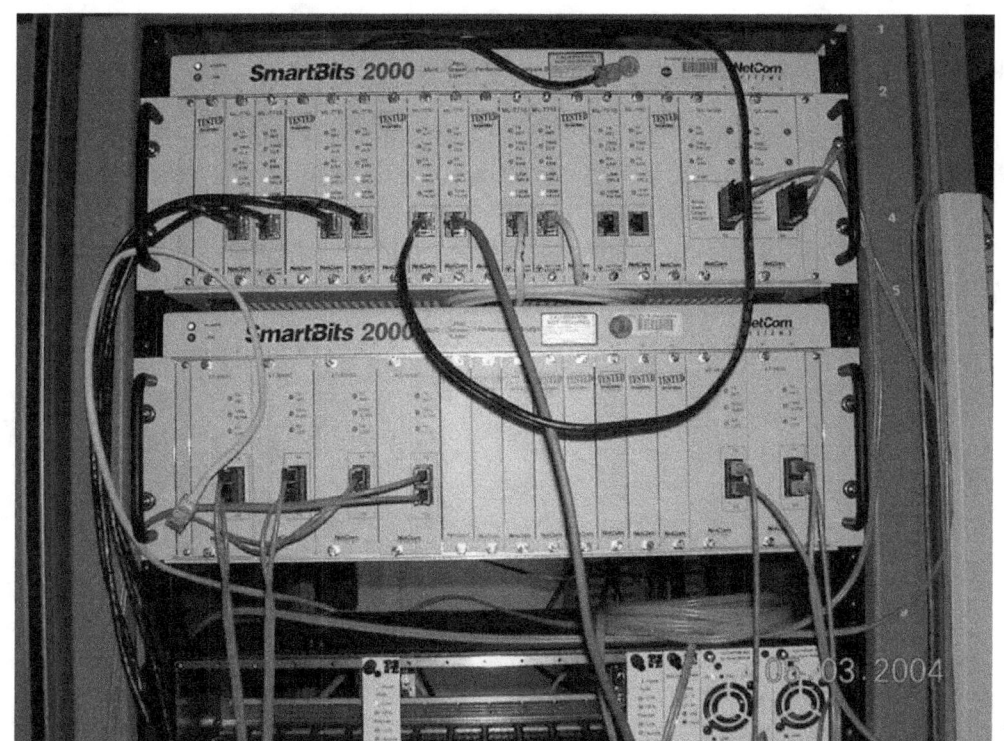

Figure 2-21**SmartBits Test Unit**

Figure 2-22**Field Testing with the SmartBits & a Laptop**

2.6.4 SONAbeam Terminal Controller Software

The main Operating System for the fSONA 622-M is the Terminal Controller Software. This was loaded on the laptops. This software allows for initialization and monitoring of the OTUs. An Ethernet connection is made from the laptop's COM1 port to an RJ-45 port within the connection box. A DB-9 to RJ-45 adapter is used on the back of the laptop. An alternate DB-9 port is also available within the connection box. Various screens of the controller software are shown in the following figures.

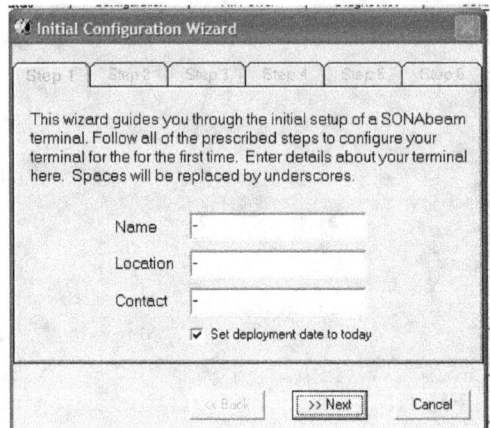

Figure 2-23   Initialization Page for fSONA Terminal Controller Software

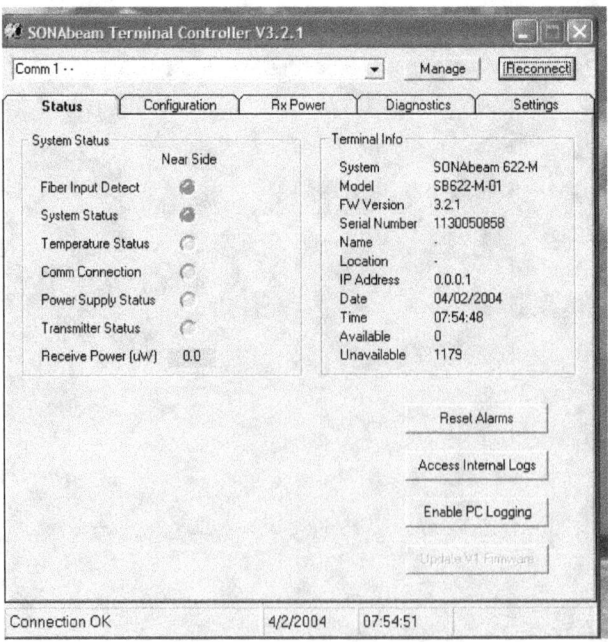

Figure 2-24   Comm 1 Status Showing No FSO Connection or Input

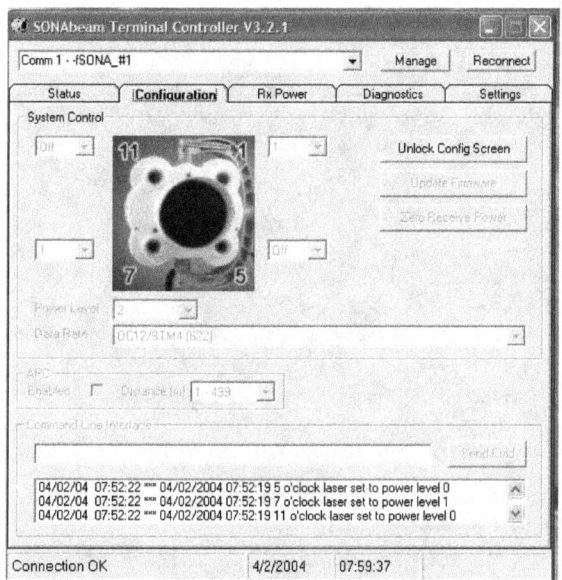

Figure 2-25   **Comm 1 Tx Configuration With System Power At Level 1**

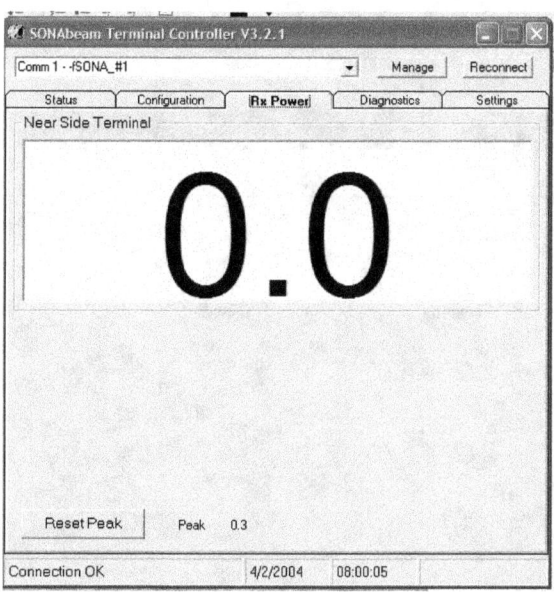

Figure 2-26   **Comm 1 Rx Power Level**

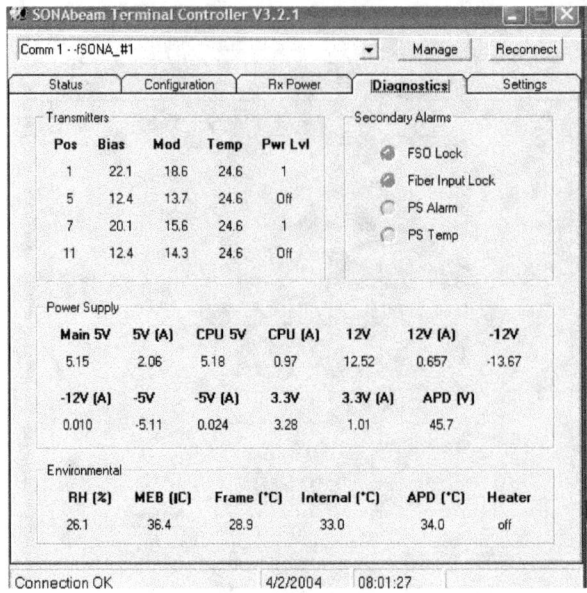

Figure 2-27   Comm 1 Diagnostics with No FSO Lock or Input

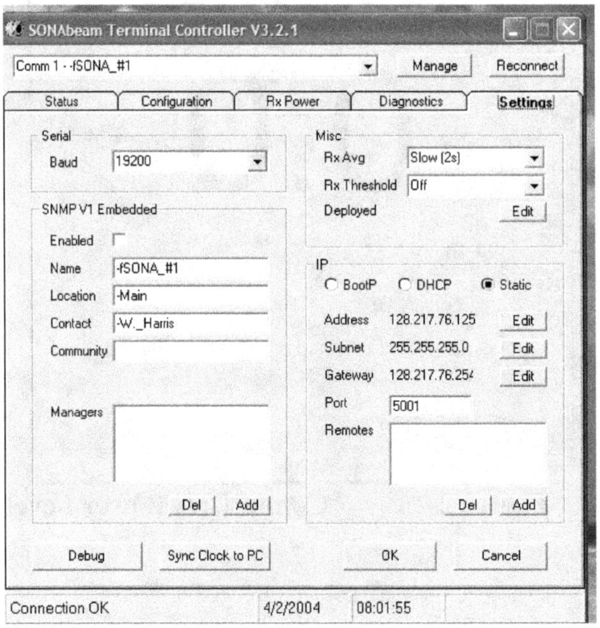

Figure 2-28   Comm 1 Settings

2.6.5 SmartApplications

SmartApplications is the operating system software for the SmartBits. This software enables the user to setup the communication links (Cards 17 to 19 for these tests) and to control the specifics of each tests. The software User Manual[6] and hardware-operating manual[7] are normally available within the ANDL.

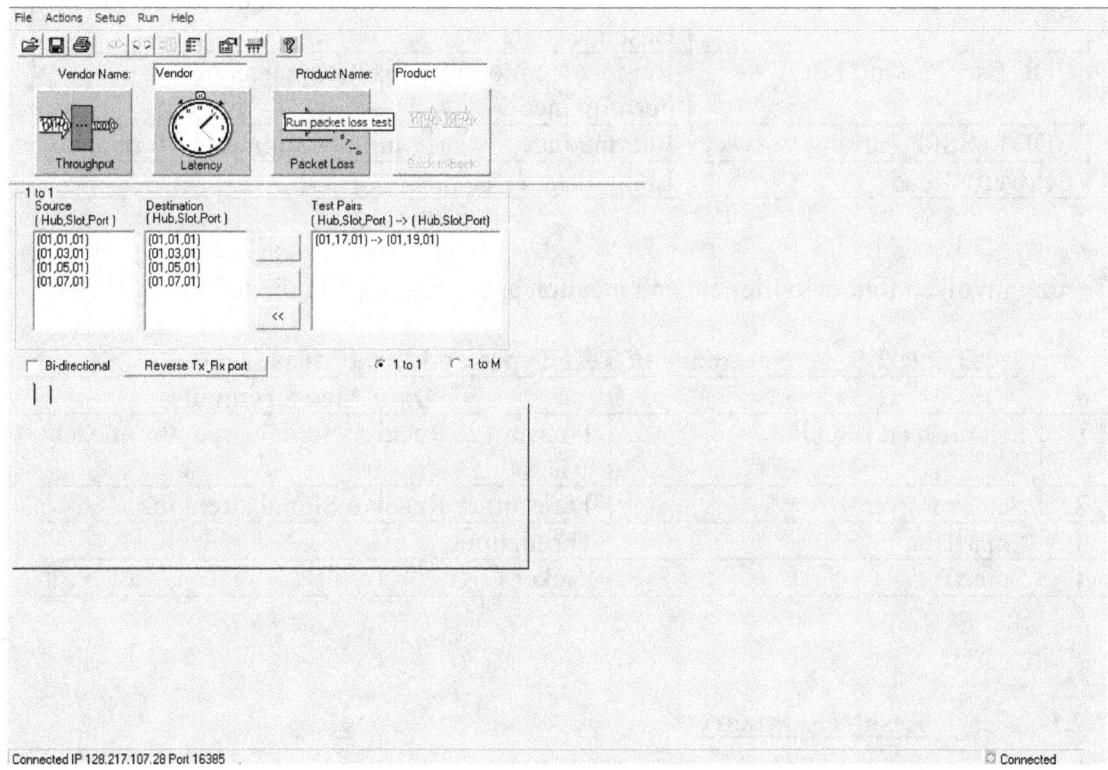

Figure 2-29   **Smart Applications Main Page with Card 17 to Card 19 Test Setup**

Figure 2-29 shows a test setup for Card 17 to input data to the OTU with Card 19 receiving the return data. For most tests, data was sent in the other direction (from Card 19 to 17). The three types of tests available, Throughput, Latency, and Packet Loss, are also shown in this figure. Only Throughput and Packet Loss tests were run on the FSO equipment. The Latency tests are not applicable to FSO laser hardware testing.

---

[6] Smart applications for Ethernet, Token Ring, ATM; User Guide; Net Com systems; 3/21/98
[7] SmartBits – Advanced Multiport Performance Tester / Simulator / Analyzer; SMB-2000; Getting Started

2.7 TEST RESULTS

FSO testing consisted of testing at four test locations with two types of tests at each location. The four locations and the general test objectives at each are summarized in the following table.

Table 2-8 **Summary of Test Locations & Objectives**

Location	Test Objectives
EDL ANDL	Setup, Initialization, Checkout, Familiarization, Baseline
EDL East Parking Lot	Remote location initialization, checkout, & performance
EDL to SSPF Parking Lots	Intermediate distance initialization & performance
Schwartz Road	Long distance testing

Testing involved four activities at each location as summarized in the following table.

Table 2-9 **Summary of Test Types and Data Measurements**

#	Test Type	Data Measurements
1	Establish an FSO link[8]	Transmit & Receive Signal strengths at each OTU
2	Set Tx Power	Transmit & Receive Signal strengths
3	SmartBits	Throughput
4	SmartBits	Packet Loss

2.7.1 Results Summary

FSO testing was conducted from March 2004 through August 2004. Additional testing was interrupted by Hurricanes Charley and Frances. During the test period, the fSONA 622-M units worked without any noticeable problems. The most difficult part of operating the OTUs was establishing the initial FSO link.

Successful links were established out to 1.5 miles. A temporary link was established at 2.5 miles, but this link quickly degraded and the link was lost. Links could not be re-established at the 2.5, 2.0 or 1.75 distance. Due to the impact of preparing for, and securing, the hardware trailers repeatedly for hurricanes, the cumulative impact of time lost did not permit continued in-depth investigations at the 1.5 mile range.

In establishing links, maximum and minimum receiver power became obvious. Established values are as follows:

---

[8] Indicated by Green System Status light on Status page of Controller SW

Table 2-10   Recommended Receiver Power Levels

Item	Value (uW)
Minimun	5.0
Maximum	260

Values less than 2 microwatts usually would not establish a link. Values greater than 260 often oversaturated the receiver and caused similar problems. The over-power level was only a problem at short distances. During lab testing, a fiberglass screen was used over the receiver lens to attenuate the signal. In addition, all but one transmitter was turned off.

During testing in the EDL parking lot, the screens were not used, but the OTUs were purposely misaligned to lower receiver power levels into acceptable dynamic operational ranges.

Whenever links were established, the systems appeared to have significant reserve power. Links out to 1.5 miles were established with the transmitters set at system level 4. System levels go from 1 to 28. System level 1 is one transmitter set to level 1. System level 2 is two transmitters set to level 1. System level 4 is four transmitters set to level 1. System values that are multiples of 4 equate to all four transmitters being on at level 1 to 7. System level 28 (max) is all transmitters on at maximum level 7. At the larger distances, System values below 4 were not investigated.

The fSONA 622M also has an Automatic Power Control (APC) feature. This non-feedback control attempts to control transmitter powers based on the incoming receiver power. Various runs were made in the APC mode and these are reported in the data. The APC feature may turn this ON or OFF. It was OFF for most tests.

All testing was done in typically good weather. Due to testing at remote locations, lightning hazards, and the potential for damage to expensive test equipment as well as injuries to ECT personnel, no testing in the rain or during unsafe hurricane wind conditions was attempted. Fog opportunities did not occur.

Figure 2-30 summarizes the minimum receiver power as a function of distance. All values were measured at Receiver #1 with Transmitter #2 set at System Level 4 or lower.

Figures 2-31 and 2-32 show typical receiver power levels at various opposite transmitter system levels for 1.0 and 1.5 miles respectively.

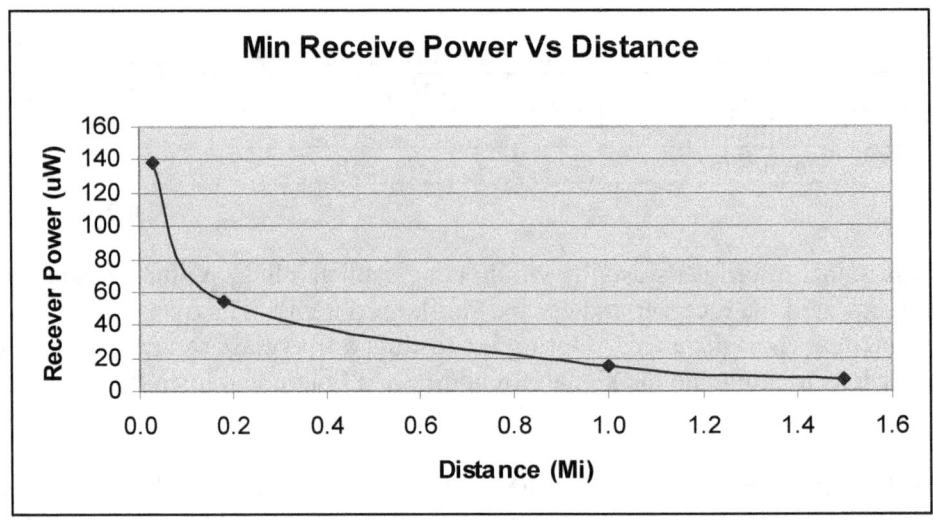

Figure 2-30   Minimum Receive Power versus Distance

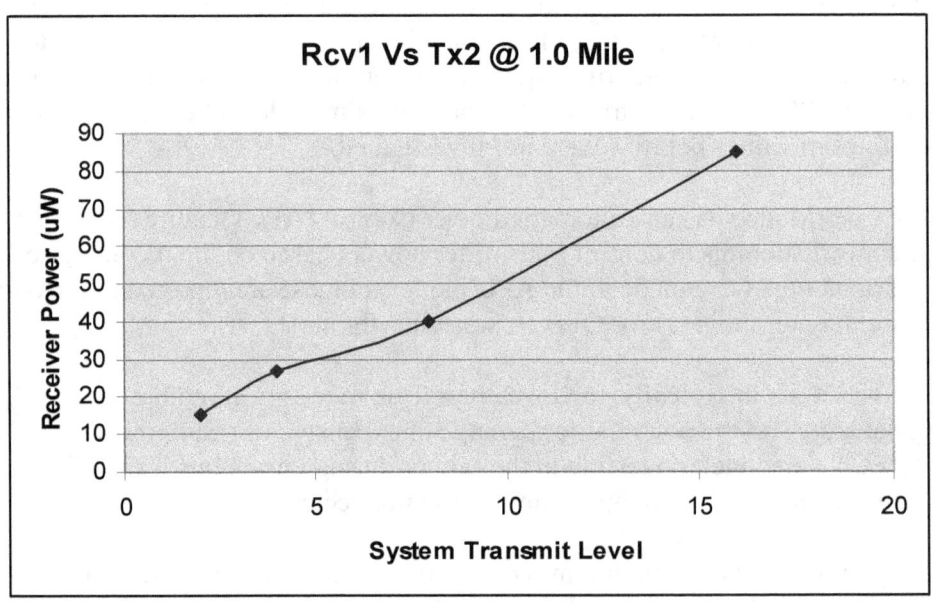

Figure 2-31   Receive Power versus Transmitter Setting @ 1.0 Mile

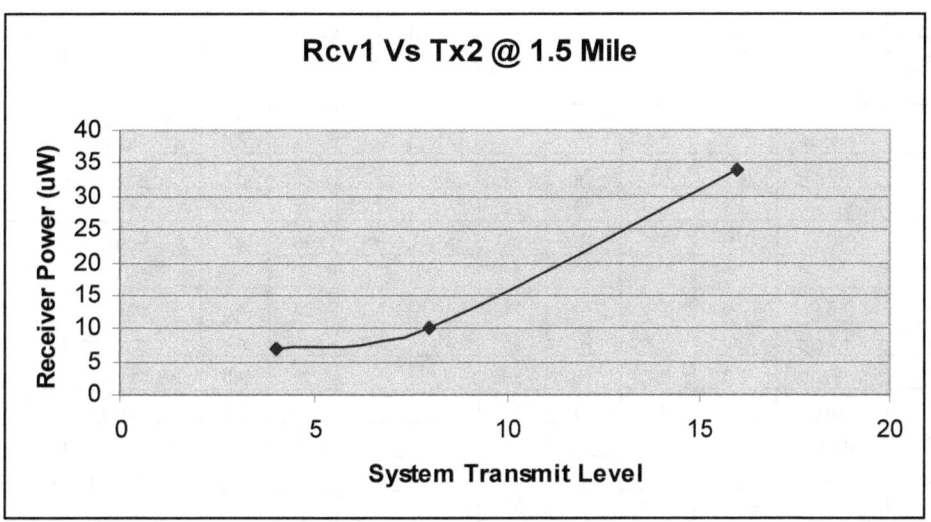

Figure 2-32   Receive Power versus Transmitter Setting @ 1.5 Mile

A summary of FSO testing is presented in the following table.

Table 2-11   Summary Of fSONA 622-M Testing

Date	Loc	Dist (ft/mi)	Unit #1 Receive uW	Attn dB	Tx Sys Lvl	Lsr On	Unit #2 Rx uW	Tx Sys Lvl	Lsr On	Comments
8/18/04	Schwartz	1.5	1.7	10	28	All	0.5	28	All	Trl #1 repositioned to 1.5 miles, System status=red
8/18/04	Schwartz	1.75	1.1	10	28	All	0.4	28	All	System status=red
8/18/04	Schwartz	1.75	1.2	10	28	All	0.4	28	All	Optimum meter alighment for #1, scope #1 down & ?, trl not in view, scope #2 down& left, System status=red
8/18/04	Schwartz	1.75	0.5	10	28	All	-	28	All	Optimum scope alighment for #1
8/18/04	Schwartz	1.75	0.0	10	28	All	0.5	28	All	Sys Status=red, same position as yesterday
8/17/04	Schwartz	1.75	1.0	10	28	All	0.5	28	All	Scope #1 up & left, Rx2 went to 0.0 with no traffic
8/12/04	Schwartz	2.0	1.0	10	28	All	0.1	28	All	Scope #2 high & far right, scope #1 down & left, Sys Status=red
8/12/04	Schwartz	2.0	0.6	10	28	All	-	0	0	Rx1=0.6 with #2 off
8/12/04	Schwartz									Secured trailers, Hur. Charley

Date	Tester									Notes
8/10/04	Schwartz	2.5	0.1	10	28	All	-	28	All	Sys Status=red
8/9/04	Schwartz	2.5	0.3	10	28	All	0.3	28	All	Rx1 decreased from 130 to 0.3, Rx2: 78 to 0.3
8/9/04	Schwartz	2.5	130	10	28	All	75	28	All	Link established
8/9/04	Schwartz	2.5	-	10	28	All	75	28	All	Moved to 2.5 miles
8/9/04	Schwartz	2.0	0.9	10	28	All	0	28	All	Sys Status=red
8/5/04	Schwartz	2.0	1.7	10	28	All	1.2	28	All	Sys Status=red
8/5/04	Schwartz	2.0	1.8	10	28	All	1.2	20	All	Sys Status=red
8/5/04	Schwartz	2.0	0.7	10	20	All	-	20	All	Sys Status=red
8/5/04	Schwartz	2.0	0.9	10	16	All	1.8	20	All	Sys Status=red
8/5/04	Schwartz	2.0	-	10	16	All	1.9	16	All	Sys Status=red
8/5/04	Schwartz	2.0	0.7	10	16	All	0.7	0	0	Rx1=0.7 with #2 off
8/3/04	Schwartz	2.0								East trailer moved
8/2/04	Schwartz	1.5	5	10	2	1&7	1.7	2	1&7	Some loss, trigger lgts blink red
8/2/04	Schwartz	1.5	5	10	2	1&7	1.7	2	1&7	Some < 100%, trigger lgts blink red
8/2/04	Schwartz	1.5	7	10	4	All	5	4	All	Zero loss
8/2/04	Schwartz	1.5	7	10	4	All	5	4	All	All 100%
8/2/04	Schwartz	1.5	10	10	8	All	6	8	All	Zero loss
8/2/04	Schwartz	1.5	10	10	8	All	6	8	All	All 100%
8/2/04	Schwartz	1.5	34	10	16	All	15	16	All	Baseline, zero loss
8/2/04	Schwartz	1.5	34	10	16	All	15	16	All	Baseline, all 100%, overcast, 89F
7/29/04	Schwartz	1.0	15	10	2	1&7	6.5	2	1&7	Zero loss
7/29/04	Schwartz	1.0	15	10	2	1&7	6.5	2	1&7	All 100%
7/29/04	Schwartz	1.0	27	10	4	All	6.5	4	All	Zero loss
7/29/04	Schwartz	1.0	27	10	4	All	6.5	4	All	All 100%
7/29/04	Schwartz	1.0	40	10	8	All	10	8	All	Zero loss
7/29/04	Schwartz	1.0	40	10	8	All	10	8	All	All 100%
7/29/04	Schwartz	1.0	85	10	16	All	25	16	All	Baseline, zero loss
7/29/04	Schwartz	1.0	85	10	16	All	25	16	All	Baseline, all 100%, partly cloudy, 90F
7/28/04	Schwartz	1.0	75	10	16	All	25	16	All	Over ride APC
7/28/04	Schwartz	1.0	75	10	16	All	25	16	All	Over ride APC
7/28/04	Schwartz	1.0	280	10	15	All	20	APC	All	Over ride APC, set distance
7/28/04	Schwartz	1.0	295	10	20	All	-	APC	All	Test failed & hung up SB operating SW
7/28/04	Schwartz	1.0	275	10	APC	1&7	2.2	APC	All	APC causing problems, trigger lights flash randomly
7/28/04	Schwartz	1.0	275	10	APC	1&7	2.2	APC	All	APC causing problems, trigger lights flash randomly
7/28/04	Schwartz	1.0	275	10	APC	All	12	APC	All	.
7/27/04	Schwartz	1.0	275	10	28	All	160	28	All	Setup at Schwartz Rd

ECT - Phase 3

Date	Location									Notes
7/26/04	Schwartz	1.0								trailers moved to Schwartz Rd
7/19/04	SSPF	950	45.2	10	APC	All	50.2	4	All	System status red, FSO lock = red, FSO input lock = red, SmartBits not connected due to rain
7/19/04	SSPF	950	45.2	10	APC	All	-	4	All	Changed focus, Trigger lights off
7/19/04	SSPF	950	69.0	10	APC	All	-	4	All	Changed focus until trigger light comes on
7/19/04	SSPF	950	12.2	10	APC	All	-	4	All	APC #2 off, APC #1 on, light rain
7/19/04	SSPF	950	12.5	10	APC	All	-	4	All	APC #2 off, APC #1 on, light rain
7/19/04	SSPF	950	-	10	APC	-	1.7	4	All	APC #2 off, APC #1 on, Rx2=1.7 max, status=green, light rain
7/19/04	SSPF	950	111.2	10	APC	-	11.3	APC	All	Tx1 APC on, System status red @ #2, light rain
7/19/04	SSPF	950	111.2	10	APC	-	-	APC	All	Tx1 APC on, RxLOS red light still on, light rain; Did not receive learning frame; recycled SmartBits
7/19/04	SSPF	950	111.2	10	APC	1&7	-	APC	All	Tx1 APC on, RxLOS red light still on, light rain; Did not receive learning frame
7/19/04	SSPF	950	580.4	10	APC	1&7	-	APC	All	Tx1 APC on, RxLOS red light still on, light rain; Did not receive learning frame
7/19/04	SSPF	950	550.0	10	APC	All	-	APC	All	Tx1 APC on, trigger red lights still on, light rain
7/19/04	SSPF	950	15.0	10	APC	All	-	APC	All	Tx1 APC on, RxLOS red light still on, light rain
7/19/04	SSPF	950	56.2	10	APC	All	-	APC	All	Tx1 APC on, RxLOS red light still on, light rain; Did not receive learning frame
7/19/04	SSPF	950	55.9	10	16	All	-	APC	All	Increase Tx1, RxLOS red light still on, light rain
7/19/04	SSPF	950	55.6	10	15	All	-	APC	All	Increase Tx1, RxLOS red light still on, light rain
7/19/04	SSPF	950	55.6	10	14	All	-	APC	All	Increase Tx1, RxLOS red light still on, light rain
7/19/04	SSPF	950	55.6	10	13	All	-	APC	All	Increase Tx1, RxLOS red light still on, light rain
7/19/04	SSPF	950	55.6	10	12	All	-	APC	All	Increase Tx1, RxLOS red light still on, light rain
7/19/04	SSPF	950	55.6	10	11	All	-	APC	All	Increase Tx1, RxLOS red light still on, light rain
7/19/04	SSPF	950	55.6	10	10	All	-	APC	All	Increase Tx1, RxLOS red light still on, light rain
7/19/04	SSPF	950	55.0	10	9	All	-	APC	All	Increase Tx1, RxLOS red light still on, light rain
7/19/04	SSPF	950	55.0	10	8	All	-	APC	All	Increase Tx1, RxLOS red light still on, light rain
7/19/04	SSPF	950	55.0	10	7	All	-	APC	All	Increase Tx1, RxLOS red light still on, light rain
7/19/04	SSPF	950	55.0	10	6	All	-	APC	All	Increase Tx1, RxLOS red light still on, light rain
7/19/04	SSPF	950	55.0	10	5	All	-	APC	All	Increase Tx1, RxLOS red light still on, light rain
7/19/04	SSPF	950	55.0	10	4	All	13.1	APC	All	Recycled SmartBits, RxLOS light on, De-focused to Rx1=55

ECT - Phase 3

Date	Site	Col3	Col4	Col5	Col6	Col7	Col8	Col9	Col10	Notes
7/19/04	SSPF	950	365.7	10	4	All	13.1	APC	All	Sys green;
7/19/04	SSPF	950	-	10	4	All	13.1	APC	All	APC #2 on; 1 & 5 @ lvl 5, 5 7 11 @ lvl 4; light rain
7/19/04	SSPF	950	-	10	4	All	3.6	16	All	Reset Tx2 to sys lvl 16 (4 @ lvl 4)
7/19/04	SSPF	950	54.6	10	4	All	2.5	4	All	Test OK
7/19/04	SSPF	950	54.6	10	4	All	2.5	4	All	Test OK
7/19/04	SSPF	950	45.0	10	4	All	-	4	All	Min Rx1 value to keep trigger lights off
7/19/04	SSPF	950	70.0	10	4	All	-	4	All	Max Rx1 value to keep trigger lights off
7/19/04	SSPF	950	64.3	10	4	All	-	4	All	Adjusted el @ #1 (East end)
7/19/04	SSPF	950	27.3	10	4	All	-	4	All	Recycled SmartBits with no change
7/14/04	SSPF	950	54.3	10	4	All	2.2	4	All	Could not adjust #2 higher than 2.2
7/14/04	SSPF	950	54.3	10	4	All	2.2	4	All	Test OK
7/14/04	SSPF	950	55.1	10	4	All	2.2	4	All	Test OK
7/14/04	SSPF	950	55.1	10	4	All	2.2	4	All	Test OK
7/14/04	SSPF	950	71.5	10	4	All	-	4	All	Test OK
7/14/04	SSPF	950	71.5	10	4	All	-	4	All	Test OK
7/14/04	SSPF	950	69.6	10	4	All	-	4	All	Lower Tx1 to sys lvl 4 (4 @ lvl 1)
7/14/04	SSPF	950	71.0	10	8	All	-	4	All	Lower Tx1 to sys lvl 8 (4 @ lvl 2)
7/14/04	SSPF	950	72.1	10	12	All	-	4	All	Lower Tx1 to sys lvl 12 (4 @ lvl 3)
7/14/04	SSPF	950	73.0	10	16	All	-	4	All	Increase Tx1 to sys lvl 16 (4 @ lvl 4)
7/14/04	SSPF	950	622.0	10	16	All	-	4	All	Increase Tx1 to sys lvl 16 (4 @ lvl 4)
7/14/04	SSPF	950	622.0	10	12	All	-	4	All	Re-focus
7/14/04	SSPF	950	180.0	10	12	All	-	4	All	Increase Tx1 to sys lvl 12 (4 @ lvl 3)
7/14/04	SSPF	950	180.0	10	8	All	-	4	All	Increase Tx1 to sys lvl 8 (4 @ lvl 2)
7/14/04		SSPF	950	176.0	10	4	All	33	4	All	Focus; #2 lasers are hot or #1 receiver is sensitive
7/14/04		SSPF	950	177.5	10	4	All	33	4	All	Focus; #2 lasers are hot or #1 receiver is sensitive
7/14/04	SSPF	950	46.0	10	4	All	19.4	4	All	De-focus
7/14/04	SSPF	950	90.0	10	4	All	19.4	4	All	De-focus
7/14/04	SSPF	950	250.5	10	4	All	19.4	4	All	De-focus
7/14/04	SSPF	950	702.0	10	4	All	19.4	4	All	Peaked arount 700 @ m#1, too hot to use; had to de-focus
7/14/04	SSPF	950	107.2	10	4	All	19.4	4	All	Adjusted to up Rx2 from 0.5 to 19.4
7/12/04	EDL PL	154	137.6	10	4	All	3.4	4	All	
7/12/04	EDL PL	154	137.6	10	4	All	3.4	4	All	Recycled SmartBits power
7/12/04	EDL PL	154	139.1	10	4	All	3.4	4	All	Defocused unit #1 to lower receive pwr
7/12/04	EDL PL	154	252.	10	4	All	3.4	4	All	Defocused unit #1 to lower

ECT - Phase 3

				3						receive pwr
7/12/04	EDL PL	154	711.3	10	4	All	646.0	4	All	System pwr increased to 4 (4 @ lvl 1)
7/12/04	EDL PL	154	711.3	10	3	1,5,7		4	All	System pwr increased to 3 (3 @ lvl 1)
7/12/04	EDL PL	154	711.3	10	2	1,7		4	All	System pwr increased to 2 (2 @ lvl 1)
7/12/04	EDL PL	154	711	10	4	All		4	All	System pwr increased to 2 (2 @ lvl 1)
7/12/04	EDL PL	154	711	10	4	All	646.0	4	All	Change both power levels to Sys 4 (4 @ lvl 1)
7/12/04	EDL PL	154	711	10	8	All	725.0	8	All	Refocus Unit #2 (South end of parking lot)
7/12/04	EDL PL	154	40.6	10	8	All	47.3	8	All	Change both power levels to Sys 8 (4 @ lvl 2)
7/12/04	EDL PL	154	57.3	10	12	All	65.9	12	All	Change both power levels to Sys 12 (4 @ lvl 3)
7/12/04	EDL PL	154	92.9	10	16	All	104	16	All	
7/12/04	EDL PL	154	92.9	10	16	All	104	16	All	Change both power levels to Sys 16 (4 @ lvl 4)
7/12/04	EDL PL	154	220	10	24	All	266.0	24	All	Baseline
7/9/04	EDL PL	154	125	10	24	All	209	24	All	
7/9/04	EDL PL	154	130	10	24	All	209	24	All	
7/9/04	EDL PL	154	141	10	24	All	201.0	24	All	
7/9/04	EDL PL	154	141	10	24	All	201.0	24	All	
7/9/04	EDL PL	154	141	10	24	All	201.0	24	All	
7/9/04	EDL PL	154	141	10	24	All	201.0	24	All	Change both power levels to Sys 24 (4 @ lvl 6)
7/9/04	EDL PL	154	25.8	10	28	All	306.0	24	All	Focus #2, stopped at 306 to avoid saturation
7/9/04	EDL PL	154	25.8	10	28	All	18.4	24	All	Focus #1
7/9/04	EDL PL	154	5.7	10	28	All	4.0	24	All	
7/9/04	EDL PL	154	5.7	10	0	0	4.0	24	All	Reflective energy off #1 or background
7/8/04	EDL PL	154	0.2	10	28	All	50.0	0	0	Reflective energy off #2 or background
7/8/04	EDL PL	154	9.6	10	28	All	50.0	24	All	Increased #2 to Sys 28 (4 @ lvl 7)
7/8/04	EDL PL	154	9.1	10	24	All	50.0	24	All	Smart Bits connected
7/8/04	EDL PL	154	2.1	10	4	All	9.6	8	All	No Smart Bits
7/8/04	EDL PL	154	2.1	10	4	All	6.5	4	All	No Smart Bits
7/8/04	EDL PL	154	2	10	4	All	0.6	4	All	No Smart Bits
7/7/04	EDL PL	154	1.5	10	4	All	0.5	4	All	No Smart Bits
7/7/04	EDL PL	154	2.2	10	spec	5	0.6	2	1,7	No Smart Bits
6/17/04										Units mounted on trailers
Attenuation screens used for most tests that follow										

ECT - Phase 3

Date	Location		Col4	Col5	Col6	Col7	Col8	Col9	Col10	Notes
6/16/04	Lab		178	10	spec	5	240.0	spec	1	Same as previous test, w/screens
6/16/04	Lab		90.8	10	spec	5	249.3	spec	1	Same as previous test
6/10/04	Lab		90.8	10	spec	5	249.3	spec	1	Same as previous test
6/10/04	Lab		90.8	10	spec	5	249.3	spec	1	Same as previous test
6/2/04	Lab		91	10	spec	5	250.0	spec	1	Same as previous test
6/2/04	Lab		91	10	spec	5	25.0	spec	1	Same as previous test
5/18/04	Lab		55.8	10	spec	5	259.0	spec	1	Same as previous test; **Recommend Max Rx = 260**
5/18/04	Lab		55.8	10	spec	5	259.0	spec	1	Same as previous test
5/7/04	Lab		55.8	10	spec	5	252.6	spec	1	Same as previous test
5/7/04	Lab		55.8	10	spec	5	253.0	spec	1	In good data zone; Decreased power at #2 by tilting #1 down
5/7/04	Lab		55.8	10	spec	5	267.6	spec	1	Transition point; Same as previous test
5/7/04	Lab		55.8	10	spec	5	267.0	spec	1	Transition point; Decreased power at #2 by tilting #1 down
5/7/04	Lab		55.8	10	spec	5	290.0	spec	1	Decreased power at #2 by tilting #1 down
5/7/04	Lab		55.8	10	spec	5	324.0	spec	1	Inc power at #2 by tilting #1 up
5/7/04	Lab		55.8	10	spec	5	260.0	spec	1	Same as previous test
5/7/04	Lab		55.8	10	spec	5	260.8	spec	1	Inc power at #2 by tilting #1 up
5/7/04	Lab		55.8	10	spec	5	197.5	spec	1	Same as previous test
5/7/04	Lab		55.8	10	spec	5	198.2	spec	1	Inc power at #2 by tilting #1 up
5/7/04	Lab		55.8	10	spec	5	118.2	spec	1	Same as previous test
5/7/04	Lab		55.8	10	spec	5	118.2	spec	1	Inc power at #2 by tilting #1 up
5/7/04	Lab		55.8	10	spec	5	81.9	spec	1	Baseline:
5/7/04	Lab		55.8	10	spec	5	81.7	spec	1	Baseline:
5/7/04	Lab		55.8	10/10	spec	5	81.2	spec	1	Add 2nd 10 dB attenuator; w/ 10 dB on #1 input & output fibers
5/7/04	Lab		55.8	10	spec	5	80.4	spec	1	Baseline:
5/7/04	Lab		55.8	10	spec	5	80.4	spec	1	Baseline: w/ 10 dB on #1 output fiber, screen on each receiver
5/4/04	Lab		55.8	10	spec	5	94.7	spec	1	w/ 10 dB on #1 output fiber
5/4/04	Lab		55.8	10	spec	5	94.5	spec	1	Reinstall 10dB on #1 output fiber
5/4/04	Lab		55.8	0	spec	5	92.0	spec	1	Remove 10dB
5/4/04	Lab		55.8	10	spec	5	92.0	spec	1	Baseline w/10dB, 2 screens
5/4/04	Lab		55.8	10	spec	5	92.0	spec	1	Baseline w/10dB, 2 screens
5/3/04	Lab		50.5	10	spec	5	90.6	spec	1	Same as last, no pkt loss

ECT - Phase 3

Date	Location		Col4	Col5	Col6	Col7	Col8	Col9	Col10	Notes
5/3/04	Lab		50.5	10	spec	5	90.6	spec	1	Refocused #1 down to get weaker signal at #2, 10 dB still in
5/3/04	Lab		128.0	10	spec	5	347	spec	1	Refocused #1 up, rx pwr increased at #2
5/3/04	Lab		147	10	spec	5	274.0	spec	1	Add screen to #1, 10 dB, screen on #2,
5/3/04	Lab		147	10	spec	5	272	spec	1	Add screen to #1, 10 dB, screen on #2,
5/3/04	Lab		239	10	spec	5	276	spec	1	Baseline, 10 dB, screen on #2,
5/3/04	Lab		239	10	spec	5	276	spec	1	Baseline, 10 dB, screen on #2,
4/30/04	Lab		239	10	spec	5	279.0	spec	1	Baseline with 10 dB & screen on #2
4/30/04	Lab		239	10	spec	5	279.0	spec	1	Baseline with 10 dB & screen on #2
4/29/04	Lab		242	10	spec	5	278.0	spec	1	Lowered pwr lvl for each end, poor performance in Packet l;oss test
4/29/04	Lab		310	10	spec	5	373.0	spec	1	Higher pwr lvl for each end, no data transferred in Packet loss test
4/29/04	Lab		310	10	spec	5	373.0	spec	1	Upped pwr lvl for each end, no data transferred in Throughput test
4/29/04	Lab		235	10	spec	5	275.0	spec	1	Packet loss test; screen on #2 rcvr, all sys lgts green; Installed 10 dB attenuator in exit of #1
4/29/04	Lab		235	10	spec	5	275.0	spec	1	Throughput test; screen on #2 rcvr, all sys lgts green; 10 dB attenuator in exit of #1
4/29/04	Lab		235	15	spec	5	275.0	spec	1	Throughput test; screen on #2 rcvr, all sys lgts green; Installed 15 dB attenuator in exit of #1
4/29/04	Lab		235	0	spec	5	275.0	spec	1	Throughput test; screen on #2 rcvr, all sys lgts green; Removed 10 dB attenuator in exit of #1
4/29/04	Lab		235	10	spec	5	275.0	spec	1	Packet loss test; screen on #2 rcvr, all sys lgts green; 10 dB attenuator in exit of #1
4/29/04	Lab		236	10	spec	5	276.0	spec	1	Throughput test; screen on #2 rcvr, all sys lgts green; 10 dB attenuator in exit of #1
4/29/04	Lab		239	10	spec	5	441	spec	1	Rx light flickering without Smartbits transmitting. 10 dB attenuator in exit of #1
					spec					
4/26/04	Lab		204	0	spec	5	484	spec	1	Removed 10 dB attenuator in; SmartBits Rx for Card 17 is out
4/26/04	Lab		204	10	spec	5	484	spec	1	Turned down power to min for of each Tx; 10 dB attenuator in; flickering green receive at SmartBits
4/26/04	Lab		267	10	spec	5	498	spec	1	Turned down power of each Tx; 10 dB attenuator in; flickering green receive at SmartBits
4/26/04	Lab		513	10	spec	5	650	spec	1	Reinstalled 10 dB attenuator at ouput of #1, test unsuccessful; nothing being received at SmartBits
4/26/04	Lab		513	0	spec	5	650	spec	1	Reset SmartBits; test failed
4/26/04	Lab		513	0	spec	5	643	spec	1	All FSO indicators green

ECT - Phase 3

Date	Location									Notes
4/22/04	Lab		491	0	spec	5	641	spec	1	Turned down #1 laser to Lvl 2, Rx pwr at #2, Sys lgt Green, no attenuation in fiber
4/22/04	Lab		491	0	spec	5	725	spec	1	Turned down #1 laser to Lvl 4, no drop in Rx pwr at #2, Sys lgt red
4/22/04	Lab		491	0	spec	5	725	spec	1	Turned down #2 laser to Lvl 4, Sys lgt red
4/22/04	Lab		711	0	0	0	65.8	spec	1	Self-Reflected energy from #2, Sys lgt red
4/22/04	Lab		711	0	spec	5	725	spec	1	Realigning, Sys lgt red
4/22/04	Lab		711	0	spec	5	725	spec	1	Realigning, Sys lgt red
4/22/04	Lab		711	0	spec	5 & 7	725	spec	1	Realigning, Sys lgt red
4/22/04	Lab		711	0	spec	7	64	spec	1	Units had been moved, Sys light red
4/15/04	Lab		74.3	0	spec	7	105	spec	1	Repeat previous successful config; both Optical head lights green, Throughput Test
4/15/04	Lab		74.8	0	spec	7	104	spec	1	Repeat previous test after resetting SmartBits; Sonet not established
4/15/04	Lab		74.8	0	spec	7	104	spec	1	Repeat previous test for Packet loss
4/15/04	Lab		74.8	0	spec	7	104	spec	1	Jumper at #1, send at #2; still card 19 to Card 17, Throughput test; part of data missing; all Optical heads lights green
4/15/04	Lab		74.9	0	spec	7	103	spec	1	Packet Loss test, both Optical units have all green lights
4/15/04	Lab		74.9	0	spec	7	103	spec	1	wiggled 17 Rx fiber & got green LED ; ThroughPut test, both Optical units have all green lights
4/15/04	Lab		75.5	0	spec	7	103	spec	1	wiggled fibers
4/15/04	Lab		75.5	0	spec	7	103	spec	1	Back to previous successful config, did not work; #2 Fiber in=red, sys=red; #1 fiber in = green, sys=red
4/15/04	Lab		74.5	0	spec	7	103	spec	1	Repeat last test with Smartbit jumper; #2 Fiber in=green, System=red; #1 Fiber in & Sys = red
4/15/04	Lab		74.6	0	spec	7	103	spec	1	Replaced fiber pair to #2, Tx 17 (#2) to 19 (#1); Fiber input #2 = green; System Stat = red; Both red @ #1
4/15/04	Lab		75.6	0	spec	7	102	spec	1	Repeat last test, 17 (#2) to 19 (#1); reversed back fibers to unit #2; Fiber input #2 = red; System Stat = red; Both red @ #1; Sonet not established
4/15/04	Lab		75.6	0	spec	7	102	spec	1	Repeat last test, 17 (#2) to 19 (#1); reversed fibers to unit #2; Fiber input #2 = red; System Stat = red; Both red @ #1; Sonet not established
4/15/04	Lab		75.6	0	spec	7	102	spec	1	Repeat last test, 17 (#2) to 19 (#1); loop at smart bits for unused fibers; Fiber input to #2 is red; some data processed
4/15/04	Lab		75.6	0	spec	7	102	spec	1	Repeat last test, switching fiber pairs for same unit; ltr to blank
4/15/04	Lab		75.6	0	spec	7	102	spec	1	Reverse direction send 17 (#2) to 19 (#1)

ECT - Phase 3

Date	Location		Col4	Col5	Col6	Col7	Col8	Col9	Col10	Notes
4/15/04	Lab		75.6	0	spec	7	102	spec	1	Rerun prev test for Packet Loss, sys state red both units, fiber input red #2
4/15/04	Lab		75.6	0	spec	7	102	spec	1	Reset SmartBits, set acceptable thruput loss at 5% from 0%, sys state red both units, fiber input red #2
4/15/04	Lab		74.3	0	spec	7	102	spec	1	Reset SmartBits, set acceptable thruput loss at 5% from 0%
4/15/04	Lab		74.3	0	spec	7	102	spec	1	Card 19 to 17 (#1 to #2), 1 to 1, ThruPut not successful
4/14/04	Lab		76.3	0	spec	7	106	spec	1	SmartBits set to Tx #1 (card 19) & Rx #2 (card 17), loop from Tx card 17 to Rx card 19; all tests failed but some data appears to be passed. Tx=6959835, Rx 6952150
4/14/04	Lab		76.3	0	spec	7	106	spec	1	Card 19 to 17 (#1 to #2); jiggle cables helped at SmartBits; switched cables around at SmartBits, passing data but failed test
4/14/04	Lab		76.2	0	spec	7	106	spec	1	Increase #1 pwr lvl to 7 (max)
4/14/04	Lab		76.2	0	spec	7	45.7	spec	1	Increase #2 pwr lvl to 7 (max)
4/14/04	Lab		38.8	0	spec	7	43.7	spec	1	Restart
4/13/04	Lab		38.8	0	spec	7	41.1	spec	1	Turn off APC at #2 and set Tx 1 O'clock to Custom, Lvl 5
4/13/04	Lab		77.8	0	spec	7	41.1	12	all	Restart with 2nd laptop at #2 position, APC on @ #2
4/12/04	Lab		77.0	0	spec	7	41.1	12	all	Inscrease pwr level of #1 to Level 5; APC on @ #2
4/12/04	Lab		77.0	0	spec	7	13.2	spec	1	Reconnect SmartBits to #2, changed APC to enable, system lights still all green. All 4 Tx turned on
4/12/04	Lab		17.4	0	spec	7	13.5	spec	1	Reconnect SmartBits to #1, system lights still all green
4/12/04	Lab		17.4	0	spec	7	13.5	spec	1	Media converter input at #1, loop at #2. Input recognized. All system lights green at #1
4/12/04	Lab		17.4	0	spec	7	13.5	spec	1	Tried media converter input at #1. Input recognized. Input at #2 from SmartBits not recognized
4/9/04	Lab		16.2	0	spec	7	13.5	spec	1	Used night vision goggles to try & see laser orientation. Did not work
4/9/04	Lab		16.2	0	spec	7	13.5	spec	1	Zeroed received power both machines
4/9/04	Lab		8.7	0	2	1 & 7	3.4	2	1 & 7	Configured SmartBits software; no change. Diagnostic says no FSO lock & no Fiber Input Lock, Fiber input detect=Red; System status=red
4/9/04	Lab		8.7	0	2	1 & 7	3.4	2	1 & 7	Removed jumper & connected SmartBits 19 to unit #1, 17 to unit #2
4/9/04	Lab		8.7	0	2	1 & 7	3.4	2	1 & 7	Adjusted elev on #2; no link established, fiber jumper at #1
4/9/04	Lab		1.9	0	2	1 & 7	3.4	2	1 & 7	No link established, fiber jumper at #1

						1 & 7			1 & 7	
4/9/04	Lab		1.4	0	2	1 & 7	3.4	2	1 & 7	No link established, fiber jumper at #1
4/8/04	Lab		1.3	0	2	1 & 7	3.4	2	1 & 7	Used video camera to try and see laser orientation. Did not work
4/8/04	Lab		1.3	0	2	1 & 7	3.4	2	1 & 7	Realigned both; #2 was reflecting off #1 and receiving its own signal
4/7/04	Lab		0	0	4	all	0	4	all	Align units
4/6/04	Lab		0	0	4	all	0	4	all	Align units
4/2/04	Lab		0	0	4	all	0	4	all	IP's set, SmartBits cable connected
3/29/04	Lab		0	0	4	all	0	4	all	#1 & #2 Powered up & initialized, IP set

2.7.2　**SmartBits**

2.7.3　**SmartBits Testing**

SmartBits testing was performed at all locations. Test profiles included Throughput and Packet loss. Throughput tests included the following frame sizes and packet rates.

Table 2-12　SmartBits Throughput Test Parameters

Frame Size	Pks/Sec
64	706415
128	470943
256	235471
512	128439
768	83107
1024	64219
1518	44150

Packet loss tests used the same frame sizes and usually resulted in zero packet losses.

Typical Throughput test results are shown in the following tables. These were the third tests on 8/2/04. The first is throughput results and the second is Packet loss.

Table 2-13   Typical SmartBits Through-Put Test

T-080204-T3

Frame	%	Pks/Sec
64	100	706415
192	100	282566
320	100	201832
448	100	141283
576	100	108679
704	100	94188
832	100	78490
960	100	67277
1088	100	61427
1216	100	54339
1344	100	48718
1472	100	45575

Table 2-14   Typical SmartBits Packet Loss Test

T-080204-P3

Frame	%	%
64	100	0.000
192	100	0.000
320	100	0.000
448	100	0.000
576	100	0.000
704	100	0.000
832	100	0.000
960	100	0.000
1088	100	0.000
1216	100	0.000
1344	100	0.000
1472	100	0.000

2.8   FSO SECURITY CONCERNS

Security for FSO links is usually achieved by controlling physical access. The FSO beam is relatively large at a distance of one mile; diverging at approximately a 2 milli-radian angle for the beam results in a divergence of 0.002 miles at one mile, or equivalently a diameter of 10.56 feet for the beam diameter at one mile. Any covert attempt to intercept the signal would require a relatively large receiver physically placed in an obvious location within the beam or very near the transmitter. Any interception attempt would be obvious. In addition, transmitters and receivers are usually physically mounted high. This would likewise make it difficult to intercept a signal covertly without detection. Any attempts to redirect the beam would normally be immediately obvious due to the loss of signal and the breaking of the communication link. Security of the connecting fibers is likewise assured by either access control or existing fiber security methods.

2.9 COMPARISON OF MULTI-BEAM AND SINGLE BEAM SYSTEMS

The fSONA multi-beam 622-M tested in ECT Phase 3 and the AirFiber 5800 single beam tested in ECT Phase 2 are very different systems to install and operate. The biggest difference is not the number of beams, but the auto-track feature that the AirFiber system incorporates. Both systems involved a significant learning curve plus the factory training included with each was extremely valuable in shortening the operational learning curves.

The fSONA 622-M units performed well during testing out to a distance of 1.5 miles. This agrees well with the factory spec of 1.6 miles. The major difficulty with the 622-M was in the initial alignment and obtaining a system lock between the two OTU. The AirFiber units were easier to align initially due to the auto-track feature; however, the long distances of the fSONA tests were never attempted with the AirFiber units. Acquiring a signal at one mile using AirFibers built-in cameras without optical zoom would have been extremely difficult.

Both systems worked well without any observed difficulties once a link was established. The AirFiber system was tested in light to moderate rain while the fSONA units were only tested in good weather.

It is not obvious that either system has a clear technical advantage over the other. From a reliability standpoint, the simpler wide-beam, non-tracking fSONA has a theoretical advantage. Likewise, the simpler, lower-cost fSONA unit likely has a price advantage when procured in large quantities. It would appear that the optimum system for future use within the unique KSC environment would include features from both the narrow-beam, auto-tracking technology of the AirFiber design, as well as feature from the simpler, more reliable fSONA unit. This hypothetical future FSO system for use at KSC would likely include the following characteristics:

- Multiple transmitters (fSONA), thereby providing redundancy in the event of a single laser failure
- Gross Auto-tracking (AirFiber had fine auto-tracking)
- Closed-loop feedback from one unit to another (AirFiber)
- Telescope for initial alignment (fSONA)
- Internal camera for real-time view (AirFiber)
- Optical zoom on internal camera (new)

2.10 FSO SUMMARY AND RECOMMENDATIONS

The precept of ECT during Phases 2 and 3 was to investigate newly-introduced FSO products while they were still in their infancy, and thereby influence the development of these developing products as early as possible, before we needed operational FSO systems on the Range at KSC. We are now much closer to fielding practical FSO systems at KSC, and we have clearly gained an in-depth understanding of the limits of this technology within the unique KSC environment. As in any industry, the state of the art continues to improve. And we have found that there are critical unique aspects of the KSC environment that force a shift from the priorities that commonly exist for FSO systems intended for urban areas.

For example, in Phase 2, we found that the large birds native to the Merritt Island National Wildlife Refuge are quite capable of blocking the narrow laser beams of auto-tracking FSO systems, thereby introducing data link interruptions. The mitigation, investigated during Phase 3, was to switch to non-tracking, widebeam FSO systems. These systems utilize multiple laser beams that diverge to more than a meter in diameter at reasonable communication distances. The small ¼-meter diameter laser beams that are used in the auto-tracking FSO systems investigated during Phase 2 are easily blocked by large birds. The larger non-tracking multiple beam systems investigated this year, in Phase 3, are largely immune to these bird-blocking outages, despite being more difficult to install a link with relative to the auto-tracking systems.. For the KSC environment, though, the reduced vulnerability to bird-blockages is an important advantage.

Hybrid multi-beam systems with auto-track are now on the market. They provide the best of both worlds, with an increased ease in establishing a link, while having the advantages of being largely immune to bird blockages. Different wave lengths are also being marketed by new entrants, and for maintaining communication during planned burning of overgrown brush and grasses, this could be an advantage. More importantly, the impacts of networking multiple FSO systems needs to be investigated, prior to fielding multiple systems, to understand better the limits for networking multiple FSO systems at KSC. ECT follow-on activities should investigate these architectural developments over the next year, prior to fielding operational FSO systems.

Another practical application that should be investigated is an FSO system with a GPS or other tracking technology to enable high-speed initial automatic acquisition of high data-rate communication with mobile objects. Initial applications could be a slow moving object such as a Shuttle recovery convoy or a rocket roll-out, where high-speed data connectivity is desired without dragging long cables that often break, which is the practice currently used for Atlas launch-vehicle processing.

We are clearly much closer to our goal of fielding high data-rate FSO systems within the KSC environment, gaining the flexibility of not having to bury fiber optic cables or dragging such cables alongside rail-mounted convoys, while still achieving higher data rates over longer distances than is possible with Wireless Ethernet.

3.0 ULTRA WIDE BAND

3.1 INTRODUCTION

Emerging Ultra Wideband (UWB) Orthogonal Frequency Division Multiplexing (OFDM) systems hold the promise of delivering wireless data at high speeds, exceeding hundreds of megabits per second over typical distances of 10 meters or less. The result of this year's UWB research was an investigation of the timing accuracies required of the timing error which can adversely affect the positional accuracies and Bit Error Rates of UWB systems. The impact is greater on BER performance than on positional accuracy performance, hence the focus in this paper is placed primarily on BER effects of timing uncertainty rather than on positional uncertainty. If communication is not achieved at all, positional awareness, were communication to exist, is of little importance.

Specifically, the effect of these timing errors can impact the achievement of Bit Error Rates on the order of magnitude of 10^{-12} or better. The dual purpose of investigating timing accuracies is therefore to assess the positional awareness as well as the Bit Error Rate performance of UWB systems needed for addressing future NASA communication needs. By understanding the practical limits of performance due to implementation problems, it thereby becomes possible to avoid overloading the correction of irreducible errors due to misaligned timing errors to a small absolute number of bits in error in real-time relative to a data rate of hundreds of megabits per second. Then, once communication is possible at low BER, positional awareness performance can be achieved.

The research approach involved managing positional awareness errors and bit error rates through identifying the effects of maximum timing synchronization error parameters. The effect of these parametric errors determine the timing accuracies required to avoid operation of communication systems within the asymptotic region of BER flaring at low BERs in the resultant BER curves. The proposed solution to these errors is to push physical layer bit error rates to below 10^{-12} before using forward error correction (FEC) codes. This way, the maximum reserve is maintained for the FEC hardware to correct for burst as well as recurring bit errors due to corrupt bits caused by other than timing synchronization errors. The effect on location accuracy is likewise minimized.

Table 3-1   List Of UWB-Related Abbreviations

Bit Error Rates	(BER)
Binary Phase Shift Keying	(BPSK)
Bandwidth	(BW)
Clock and data recovery	(CDR)
Energy per bit	(Eb)
Federal Communications Commission	(FCC)
Forward error correction	(FEC)
Hertz	(Hz)
Inverse Fast Fourier Transform	(IFFT)
Multi-Band OFDM Alliance	(MBOA)
Mega Bits per second	(Mbps)
Multiplexer	(Mux)
Effective noise	(Neff)
Orthogonal Frequency Division Multiplexing	(OFDM)
On/Off Keying	(OOK)
Pulse Amplitude Modulation	(PAM)
Performance Analysis Tool	(PAT)
Probability density function	(pdf)
Physical	(PHY)
Pulse Position Modulation	(PPM)
Radio frequency	(RF)
Unit Interval	(UI)
Ultra Wideband	(UWB)
Wireless personal area networking	(WPAN)

3.2 BACKGROUND

In the near future, wireless broadband communications systems will require data rates exceeding hundreds of mega bits per second (Mbps). To address these approaching demands, emerging Ultra Wideband (UWB) Orthogonal Frequency Division Multiplexing (OFDM) offers an ideal physical (PHY) layer solution to address wireless personal area networking (WPAN) needs over short ranges. As UWB modulation becomes better understood and data rates increase to near their high data rate potentials, the control of timing synchronization errors will become ever more critical in measuring UWB's system performance parameters. The ECT Phase 3 research explored the timing accuracies required to support the operation of UWB OFDM systems in such a future communication landscape.

3.3 OFDM OVERVIEW

Instead of using the traditional Pulse Position Modulation (PPM), Pulse Amplitude Modulation (PAM), Binary Phase Shift Keying (BPSK), and On/Off Keying (OOK) modulations investigated in the prior two years research, an alternate approach for modulating Ultra Wideband (UWB) Pulses is investigated in this year's research, consistent with changing trends monitored in the trade press for implementing UWB communication techniques. This newer modulation technique is achieved through OFDM (Orthogonal Frequency Division Multiplexing). OFDM is a modulation technique suitable for high data rate systems for networked applications, and is more suited to future congested communications arenas than the simpler PPM, PAM, BPSK, and OOK methods of modulation.

The prominent emerging UWB OFDM system has been developed by the Multi-Band OFDM Alliance (MBOA http://www.mboaalliance.com) and uses the OFDM technique to occupy the statutory wide bandwidths permitted for UWB systems. The basic idea of OFDM involves splitting a high-rate data stream X_N into a number of lower rate streams that are transmitted simultaneously at different frequencies over a number of sub-carriers ($X_0, X_1, \ldots X_{N-1}$) (See Figure 3-1).

Figure 3-1   De-Multiplexed High Data Rate Of DM Data Stream

In OFDM the sub-carrier pulse used for transmission is a rectangular pulse. With this rectangular pulse, the task of pulse forming and modulation can be simply implemented with an Inverse Fast Fourier Transform (IFFT). According to the Fourier Transform Theorem, the rectangular pulse shaped sub-banded pulses in OFDM will lead to the [sin(x)/x] spectrum at the receiver, when the signal is Muxed back together, seen in Figure 3-2. To obtain high spectral efficiency, the frequency response of the sub-banded are overlapped and orthogonal, which means that where the signal is evaluated (at the maximum peak), the value of all other signals are zero.

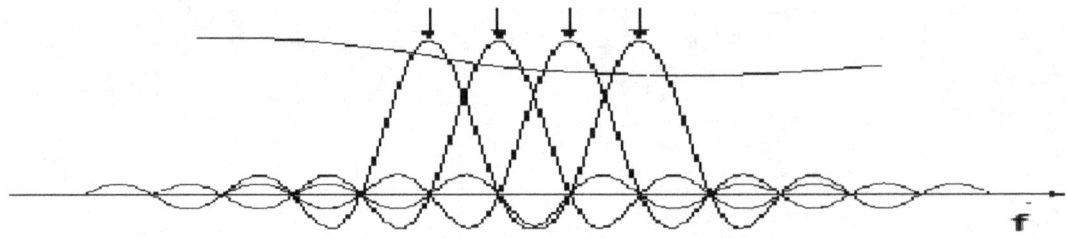

Figure 3-2   Overlapping Orthogonal Sub-Carriers In OFDM Symbol

Major benefits of the OFDM technique include higher spectral efficiency, resiliency to radio frequency (RF) interference, and lower multi-path distortion [4]. On the other hand, OFDM shortcomings evolve through its high sensitivity to frequency and time synchronization error compared to single carrier system [4]. Frequency synchronization error results from misalignment in sub-carrier frequencies due to fluctuations in radio frequency oscillators or channel's Doppler frequency introducing inter carrier interference (ICI). Timing synchronization errors refer to the incorrect timing of the OFDM symbols at the demodulator introducing inter symbol interference (ISI). [3]. Both ICI and ISI cause bit errors in a UWB-OFDM system. The focus of this research was to analyze how accurate timing synchronization errors must be to obtain a bit error rate of 10^{-12}, or better, assuming perfect frequency synchronization. A closer examination of this phenomenon will be shown in Section 2.

Although currently there are two major UWB proposals, consisting of single band and multi-band impulse-centered approaches, both with their own advocates fighting for their approach to become the accepted IEEE and FCC standard, in this paper we focus on the multi-banded OFDM approach and its concepts and limitations. This focus is chosen since the multi-band approach currently has achieved more favor among the candidate approach to become the universal UWB standard.

3.4 RESEARCH ORGANIZATION

The importance of controlling timing synchronization errors was established earlier. Discussions regarding timing synchronization errors in UWB-OFDM systems are discussed starting the next section. Then starting in Section 3.10, new methodologies are presented for analytical solutions for bit synchronization effects on BER (Bit Error Rate). Starting in Section 3-20, results from analytical and experimental analyses are presented along with a hypothesis of the timing error effects. Finally, in Section 3-27 the research concludes by restating the research objectives, documenting why timing synchronization is important, and summarizing the effect on BER and positional accuracy performance if the timing accuracies are not maintained in a UWB-OFDM data link.

3.5 TIMING JITTER IN UWB-OFDM COMMUNICATION SYSTEMS

Measured performance of a digital data transmission system usually is obtained through analyzing the probability of error at a given bit error rate and signal-to-noise ratio. As the UWB systems evolve into their expected achievable high data rate values, controlling timing synchronization errors becomes essential since timing errors cause bit errors that degrade UWB-OFDM system performance. To present the concept of "How timing errors affect UWB-OFDM system performance", this section is divided into three parts. First the OFDM symbol structure is described. Then, how timing errors in the OFDM symbols affect the system's performance is discussed; and third is an analysis of the impact of timing jitter in digital communication systems. This section concludes with a Tikhonov approximation approach for estimating the timing error.

3.6 OFDM TECHNIQUE

Orthogonal Frequency Division Multiplexing (OFDM) is a flexible technique that increases bandwidth efficiency, resiliency to radio frequency (RF) interference, and lower multi-path distortion. For example, if interference with an existing narrowband system occurs, UWB -OFDM permits simply not using one or more particular sub-bands. This technique can be thought of as analogous to a combination of multi-carrier modulation (MCM) and frequency shift keying (FSK). MCM divides a data stream into several bit streams and modulates each bit stream onto sub-carriers [16]. FSK transmits data onto one carrier from multiple orthogonal carriers. Orthogonality between the sub-bands among an UWB-OFDM modulation format is accomplished by separating the bands by an integer multiple of the inverse of symbol duration of the parallel bit streams [4] Orthogonality in the symbol is crucial because it helps to eliminate inter-symbol interference (ISI) and inter-carrier interference (ICI). This is best done by adding a guard time insertion or a cyclic prefix to the beginning of the OFDM symbol. Cyclic prefix (CP) involves attaching a copy of the last part of the OFDM symbol to the beginning of the symbol as shown in Figure 3-3.

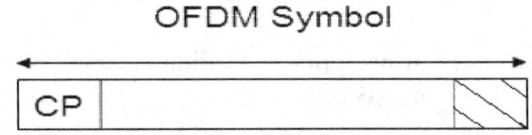

Figure 3-3   OFDM Symbol

In the transmitter, after the parallel data of N sub-channels are modulated onto N sub-carriers ($d_0, d_1, \ldots d_{N-1}$), where each d_N represents a complex number, they are fed into an Inverse Fast Fourier Transform (IFFT). The transmitted data is given by [5]:

$$s(t) := \sum_{k=-\infty}^{\infty} \sum_{i=0}^{N-1} d_i(k) \exp\left[j2\pi f_i(t - kT_s)\right] f(t - kT_s)$$

where T_s is the symbol duration of the OFDM pulse and f_i ($i=0.1,\ldots N-1$) is the frequency of the *i*th sub-carrier given by [5]:

$$f_i := f_0 + \frac{i}{T_s}$$

Here, f(t) is the pulse waveform of each of the symbols and it is defined as

$$f(t) = \begin{cases} 1 & (0 \leq t \leq T_s) \\ 0 & \text{(otherwise)} \end{cases}$$

After the IFFT process, the signal s(t) goes through a guard time insertion circuit where the cyclic period is added so inter-symbol interference can be avoided as much as possible.

One requirement of adding a cyclic prefix to the symbol is that it should be longer than the impulse response of the channel. When the cyclic prefix to the symbol is longer than the impulse response of the channel, it acts as a guard space between the sub-carriers. The guard time is chosen larger than the expected delay spread, such that multi-path components from one symbol cannot interfere with the next symbol [15]. This eliminates ISI and ICI almost completely. However, some residual ICI may still exist. This happens when the multi-path delay becomes larger than the guard time. At this point, the system may manifest timing errors due to the cumulative effects of multi-path delay variations.

At the output of the guard time insertion circuit, the OFDM symbol is given by:

$$\frac{d}{dt}s(t) := \sum_{k=-\infty}^{\infty} \sum_{i=0}^{N-1} d_i(k) \exp\left[j2\pi f_i (t - kT_{total})\right] \frac{d}{dt}f(t - kT_{total})$$

[5]

where the modified pulse waveform of each symbol is defined as

$$\frac{d}{dt}f(t) = \begin{vmatrix} 1 & (-T_g \leq t \leq T_s) \\ 0 & (t < -T_g, t > T_s) \end{vmatrix}$$

[5]

3.7   TIMING ERROR EFFECTS ON SYSTEM PERFORMANCE

Timing signals play several of different roles in communication systems. One example evolves in digital systems, where clock signals are used to transfer logic signals in and out of registers at times when their values are valid. The maximum clock frequency is usually limited by the propagation delay of the logic circuits between registers. In high bandwidth digital input/output systems, however, the date transfer rate can be limited by uncertainty in the clocks used to transfer the data [21]. Fixed offsets between transmit and receive clocks or timing errors due to noise comprise this uncertainty.

Timing errors in UWB-OFDM are simply a short variation of the OFDM sub-band's bit timing from its ideal time slot location. A bit's timing is simply the composite effect of multiple monocycles acting in concert due to various multi-path delays acting in addition to the composite effect of digital circuitry timing errors. Effects of errors in the time base of the signal, due to timing errors can also limit performance parameters such as achievable bit rates of the system. When this happens the system is said to have a degraded performance due to timing errors.

3.8   TIMING JITTER AND PHASE NOISE RELATIONSHIP

Oscillator or clock uncertainties in synchronous digital systems can degrade a system's performance, resulting from in bit errors. Phase noise and timing jitter result from uncertainties in the clock's oscillator output. Phase noise defines the frequency output of the oscillator. For example, when the output to a noisy phase oscillator is given by:

$$V(t) = V_o \cdot \cos\left[\omega_o(t) + \phi(t)\right]$$

[17]

then

ϕ(t)

is the phase noise also referred to as the spectral density of phase fluctuation. The random fluctuations of phase, that are responsible for phase noise, can also be observed in the time domain as timing jitter. Given that timing jitter is a measure of variation in the time domain, it ultimately describes how far a bit period wanders from its ideal location. In OFDM systems, controlling timing jitter in the sub-carriers calls for precise synchronization at the OFDM demodulator. This control involves determination of the starting sample of the ith OFDM symbol such that the cyclic prefix (CP) can be disregarded and the OFDM symbol can be properly realigned [3]. For example, consider the OFDM block diagram shown in Figure 3-4. Before the OFDM symbol can be multiplexed back together and the orthogonality of the symbol at the receiver is preserved, timing errors must be controlled or even corrected to some degree. Otherwise, the system will experience inter-carrier interference (ICI). ICI is crosstalk between different sub-carriers, which means that the sub-carriers are no longer orthogonal in signal space [15]. The orthogonality of the sub-carriers can be maintained and individual sub-carriers can be separated by using an FFT (Fast Fourier Transform) circuit when there is no inter-symbol interference (ISI) and inter-carrier interference introduced by transmission channel distortion [1]. In reality, these conditions cannot be obtained. To manage distortion caused by the transmission channel, in the next section we introduce a method for estimating the timing accuracy required for achieving a given bit error rate performance for a high data rate UWB-OFDM digital link.

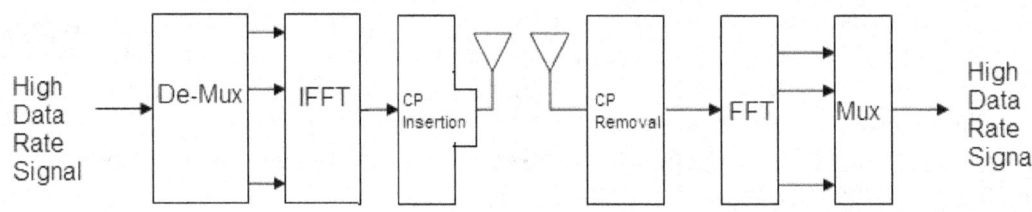

Figure 3-4   OFDM Block Diagram

3.9   TIKHONOV APPROXIMATION OF TIMING ERROR

In OFDM bit symbols there will always be some fluctuation in the bit symbol's ideal timing. This fluctuation can be estimated around the ideal timing's mean value. Since the timing errors are random values, to estimate fluctuation in timing, we characterize the actual timing with a probability density function (pdf). The pdf shows how the actual bit timing estimate can be before or after the ideal value.

In the literature [3,18,19.20], it is found that timing errors are usually characterized with a Gaussian or a Tikhonov pdf. However, the choice of a Tikhonov pdf is selected in this research so that we can obtain a more representative characterization of the bit synchronizer statistical properties in agreement with observed properties. The next section displays the use of the Tikhonov approximation of timing errors.

3.10   BIT SYNCHRONIZATION

This section introduces and presents methods and procedures for estimating the timing accuracy required for achieving a given bit error rate performance for a high data rate UWB-OFDM digital link. Through this analysis, equations can be developed to determine and identify maximum timing synchronization errors of high data rate links incorporating Manchester (Bi-Phase), Miller, RZ, or NRZ coded data.

Most high speed communication systems have a low tolerance for bit errors; the allowable uncorrected bit error rate (BER) for such systems operating in the hundreds of mega bits per second typically must fall between 10^{-9} and 10^{-12} to prevent introducing error correction overload within the error correction hardware. This means that the BER impacts of synchronization and timing errors must be analyzed to estimate the timing accuracies required to avoid overloading the correction of irreducible errors due to misaligned timing errors.

In previous studies [3,18], timing error analyses have been investigated performance in bit error regions ranging from 10^{-3} to 10^{-6}. Such a lower-performance BER is entirely appropriate for low speed communication systems operating at data rates only in the tens of mega bits per second. Unfortunately, these previously investigated regions are not sufficient for proposed high speed UWB systems having data rates in the hundreds of mega bits per second. For example, consider a system running at 500 Mbps, with a target bit error rate of 10^{-6}. Such an error rate would produce 500 bit errors per second, in such a high-speed system, and a 10^{-6} level of bit error rate would cause severe system performance degradation.

Instead of battling with the high bit error rates mentioned above, the recommended approach involves reducing bit error rates to lower rates through first identifying the performance requirements associated with, and then managing, the maximum range of timing synchronization errors. Thus, it became a research goal to determine how accurate the timing errors must be among the multitude of OFDM data streams, to avoid operation in the asymptotic region where BER flaring occurs at low BERs in the resultant BER curves. Otherwise, it becomes impossible to achieve high positional awareness accuracies and low Bit Error Rates. The approach chosen is push bit errors to below 10^{-12} before taking advantage of forward error correction (FEC) codes. This way, the maximum reserve is maintained for the FEC hardware to correct for bit errors caused by other than timing synchronization errors.

For the purpose of presentation, this section is divided into two sections: analytical solutions for bit synchronization errors and experimental test cases with PulsON 200 evaluation kit hardware manufactured by Time Domain Corporation of Huntsville, AL.

3.11 ANALYTICAL SOLUTIONS FOR BER PERFORMANCE

At the beginning of the analytical solutions research, the focus was placed on estimating a maximum achievable bit rate, conditioned on timing synchronization bit errors. Through this analysis, timing error effects on bit error rate performance of a high data rate link could be identified.

The chosen methodology closely follows the derivations and methods of Lindsey and Simon [2]. The difference between the methods in this paper and the methods discussed in [2] are that we characterize the synchronization error λ to be a normalized timing error resulting from a delay locked loop. Additionally, we examine bit synchronization for a high data rate stream instead of for symbol synchronization for narrowband applications. Also, we expand the average error probability Pe to below 10^{-12}, since one must target this region of interest for applicability to high data rate links. Finally, the standard deviation σ_λ and variance σ_λ^2 of the normalized timing error represent ratios of a normalized Unit Interval (UI) of a data bit time period in a data link. Through utilizing a Unit Interval approach, in place of an absolute timing approach, the results of this research can easily be applied to ever-increasing data rates for future, yet unimagined OFDM-UWB data links, thereby increasing the value of the research documented in this paper.

In the following section, the methodology for the analytical solutions that estimated a maximum achievable bit rate, conditioned on bit timing synchronization errors are discussed. The analysis begins with obtaining conditional error probability values of the correlation detector conditioned on a timing error. Next, an explanation of how to average the obtained conditional error probabilities over a Tikhonov pdf (probability density function) to arrive at estimations of probability of bit error at the receiver is presented.

3.12 CONDITIONAL ERROR PROBABILITIES

Given that the optimum detector for a known signal is a cross-correlator, the first step in performing the analysis is to obtain derivations for the error probability of the correlation detector conditioned on a bit synchronization error for Manchester, NRZ, RZ and Miller coded data.
Below are the equations, taken from prior literature, that is used in this analysis to obtain the conditional error probabilities for Manchester, NRZ, RZ, and Miller coded UWB-OFDM data. (See [2] for derivations.)

3.13 MANCHESTER CODED DATA

For a Manchester coded bit stream, the conditional error probability can be realized through the following equation (see Table 3-2 for variable definitions):

$$Pe(\lambda) = \frac{1}{4} \cdot \text{erfc}\left[\sqrt{\frac{Eb}{No}} \cdot (1 - 2 \cdot |\lambda|)\right] + \frac{1}{4} \cdot \text{erfc}\left[\sqrt{\frac{Eb}{No}} \cdot (1 - 4 \cdot |\lambda|)\right]$$

$$|\lambda| \leq \frac{1}{4} \tag{3-1}$$

(Where the maximum random value for timing error λ is defined to be $\frac{1}{4}$ for Manchester coded UWB-OFDM data.)

3.14 NRZ CODED DATA

Furthermore, when the bit stream is coded by NRZ data, then the following equation can be used to achieve conditional error probability values:

$$Pe(\lambda) = \frac{1}{4} \cdot \text{erfc}\left[\sqrt{\frac{Eb}{No}} \cdot (1 - 2 \cdot |\lambda|)\right] + \frac{1}{4} \cdot \text{erfc}\left[\sqrt{\frac{Eb}{No}} \cdot (1 - 4 \cdot |\lambda|)\right]$$

$$|\lambda| \leq \frac{1}{2} \tag{3-2}$$

(Where the maximum random value for timing error λ is defined to be $\frac{1}{2}$ for NRZ data.

3.15 RZ CODED DATA

Next, when the bit stream incorporates RZ coded data, the conditional error probabilities can be obtained with equation (3-3).

$$Pe(\lambda) = \frac{1}{4} \cdot \text{erfc}\left[\sqrt{\frac{\left(\frac{Eb}{No}\right)}{2}}\right] + \frac{1}{4} \cdot \text{erfc}\left[\sqrt{\frac{\left(\frac{Eb}{No}\right)}{2} \cdot ((1 - 4 \cdot |\lambda|))}\right]$$

$$|\lambda| \leq \frac{1}{4}$$

(3-3)

(Where the maximum random value for timing error is defined to be $\frac{1}{4}$ for RZ coded data.)

3.16   MILLER CODED DATA

Finally, in the event that Miller coded data is used in the bit stream, equation (3-4) represents an equation that obtains its conditional probability error values.

$$Pe(\lambda) = \frac{1}{2} + \frac{3}{16} \cdot \left(\text{erf}(a)^2\right) + \frac{1}{16} \cdot (\text{erf}(b))^2$$

$$- \frac{3}{4\sqrt{\pi}} \cdot \int_0^{\sqrt{2} \cdot c} \text{erf}(x) \cdot \exp\left[-\left(x - \sqrt{2} \cdot b\right)^2\right] dx$$

$$- \frac{3}{4\sqrt{\pi}} \cdot \int_0^{\sqrt{2} \cdot c} \text{erf}(x) \cdot \exp\left[-\left(x - \sqrt{2} \cdot a\right)^2\right] dx$$

$$|\lambda| \leq \frac{1}{4}$$

(3-4)

where

$$a := \sqrt{\frac{\left(\frac{Eb}{No}\right)}{2} \cdot (1 - 4 \cdot |\lambda|)} \qquad b := \sqrt{\frac{\left(\frac{Eb}{No}\right)}{2}} \qquad c := \sqrt{\frac{\left(\frac{Eb}{No}\right)}{2} \cdot (1 - 2|\lambda|)}$$

(And where the maximum random value for timing error λ is defined to be $\frac{1}{4}$ for Miller coded data.)

3.17 ESTABLISHING AVG. ERROR PROB. AT THE RECEIVER

Next in the analysis, after obtaining the stated conditional error probabilities in equations (3-1), (3-2), (3-3), and (3-4), they are averaged over the probability density function (pdf) $p(\lambda)$ of the bit synchronization errors, to obtain the average error probability Pe at the receiver. Average error probabilities at the receiver can be determined by equation (3-5). This equation states that the average error probability at the receiver Pe is equal to the conditional error probability $Pe(\lambda)$ averaged over the pdf $p(\lambda)$ of the normalized timing synchronization error λ.

$$Pe := \int_{-\lambda_{max}}^{\lambda_{max}} p(\lambda) \cdot (Pe(\lambda))\, d\lambda \tag{3-5}$$

Hence, λ_{max} reflect the maximum value of which is defined for in the corresponding $Pe(\lambda)$ equations in (3-1), (3-2), (3-3), (3-4), depending on the data format chosen, and $p(\lambda)$ represents a probability distribution of the normalized timing synchronization error.

Although in [5], timing synchronization error postulate a Gaussian approximation, we assume a Tikhonov pdf $p(\lambda)$ since, as stated earlier, this is a more typical characterization of timing errors observed in practical bit synchronizers. [18] Completely characterized in terms of its variance σ_λ^2 of the normalized timing error, the Tikhonov pdf for the various data formats are stated in equations (3-6) and (3-7).

For NRZ data formats, $p(\lambda)$ can be characterized by

$$p(\lambda) := \frac{\exp\left[\dfrac{\cos\cdot 2\cdot\pi\cdot(\lambda)}{(2\cdot\pi\cdot\sigma_\lambda)^2}\right]}{I0\left[\left(\dfrac{1}{2\cdot\pi\cdot\sigma_\lambda}\right)^2\right]} \quad |\lambda| \le \tfrac{1}{2} \tag{3-6}$$

when a Tikhonov pdf is assumed.
Likewise, when employing Manchester, Miller or RZ coding, all of which are base-band techniques that utilize transitions in the middle of the symbol interval, $p(\lambda)$ can be characterized by the Tikhonov pdf in equation (3-7)

$$p(\lambda) := \frac{2\exp\left[\dfrac{\cos\cdot 4\cdot\pi\cdot(\lambda)}{(4\cdot\pi\cdot\sigma_\lambda)^2}\right]}{I0\left[\left(\dfrac{1}{4\cdot\pi\cdot\sigma_\lambda}\right)^2\right]} \quad |\lambda| \le \tfrac{1}{4} \tag{3-7}$$

Table 3-2 Analytical Equation Definitions

Equation Definitions

Statistical Parameters	Units	Domain or Range
$p(\lambda) :=$ Tikhonov_pdf		
$\lambda :=$ Timing_Syn_Error	Unit Intervals (UI)	random variable
EbNo := Energy_Per Bit_to_Noise Ratio	dB	0, 1, 2…18
λ_{max} (Miller) := max_λ_value		1/4
λ_{max} (NRZ) := max_λ_value		1/2
λ_{max} (Bi – Polar) := max_λ_value		1/4
λ_{max} (RZ) := max_λ_value		1/4
$P_E :=$ Avg_Error_Prob		
$Pe(\lambda) :=$ Conditional_Error_Prob		
$\sigma_\lambda :=$ stand_deviat_of_UI	Unit Intervals (UI)	Varies
$(\sigma_\lambda)^2 :=$ variance_of_UI	Unit Intervals (UI)	Varies

By substituting (3-6) or (3-7) into (3-5) along with the Pe(λ) of the reflecting equations of the different data formats, equation (3-5) is used to generate MathCad plots, developed in Figures 3-9 through 3-12. These graphs display average probability of error versus the ratio of the energy per bit (Eb) versus the spectral noise density (No) (EbNo) for the selected UI parameters of normalized timing errors.

With the process described above, it becomes possible to obtain the attainable bit rate for a data link conditioned on normalized timing synchronization errors in Manchester, NRZ, RZ, and Miller coded data. The MathCad graphed results, in Figures 3-9, 3-10, 3-11, and 3-12, demonstrate how accurate Unit Interval (UI) normalized timing error variances must be to avoid entering the asymptotic region where BER flaring occurs. At the instance where the curves begin to flare, we identify the maximum timing synchronization bit errors that can be permitted to exist in the physical implementation of the hardware for the data link before the use of forward error correction coding.

Since this research proposed pushing bit errors to below 10^{-12} before using forward error correction (FEC) codes, as summarized through the average error probability graphs, and established with equation (3-5), one can determine how accurate timing error instances must be to achieve a bit error rate of 10^{-12}.

Later in the analysis, the average probability of error results conditioned on timing synchronization errors is used to make empirical comparisons with PulsON 200 experimental results. But first, in the following section, the research methodology is continued by comparing PulsON 200 statistical data with its theoretical performance curve.

3.18 TEST CASE WITH THE PULSON200 RADIOS

Performance Analysis Tool (PAT) software was used in conjunction with two PulsON UWB Evaluation Kit (EVK) transceivers, to gather statistical information about wireless data passed between the two radios. With the receipt this year of two PAT software upgrades, with the latest version including for the first time Eb/No estimates as well as Bit Error Rate, it became possible for the first time to assess the implementation loss associated with the PulsON radio implementations, unlike in last year's research. This PAT software is provided with the EVK transceivers to permit easily assessing link performance. With PAT, the following radio data link statistical parameters were measured:

Receiver Statistics:
- Bit error rate (BER)
- Number of bit errors
- Number of bits received
- Number of packets received
- Number of packets dropped
- Effective data rate
- Time (in seconds) that the radio has been running
- Percentage of packets received
- Temperature of the PulsON 200 radio development Module
- Energy per bit (Eb)
- Effective noise (Neff), Energy per bit to effective noise strength (Eb/Neff)

Transmitter Statistics:
- Number of transmitted bits
- Number of transmitted packets
- Time (in seconds) that radio has been running
- Temperature of PulsON 200 radio development Module

3.19   TEST PROCEDURES

ECT Phase 3 testing investigated Range vs. Data Rate, which tested the throughput BER over various distances of the UWB transceivers. After configuring the radios for establishing a simple link as outlined in [5], we proceeded with the following steps:

Step 1: Double click on the PAT icon displayed on the laptop or PC monitor

Step 2: When the GUI, similar to Fig.3-1, appears on the monitor, select appropriate Radio IP address from the pull down menu and then click the connect button.

Step 3: Once a message appear in the message area that says, "Connected to Radio," select radio mode, select link rate, and the Eb/Neff mode box from the tabbed form field.

Repeat Steps 1-3 for both radios

Step 4: After a connection has been established for both radios, use measuring tape to separate radios to the desired distance.

Step 5: Next, after performing calibration tests as specified in [5], click start radio on the transmitter radio followed by clicking start on the receiving radio.

Step 6: Analyze statistical data in the statistics frame area, paying close attention to the receiver percentage rate.

Step 7: Vary the gain in the Tabbed Form field area until the receiver percentage rate is 98% or higher. (When radios are far apart, the VGA and the threshold constant, located in the Acquisition Tab, may need to be varied to obtain receiver percentage rate of 98% or higher)

Step 8: Once receiver rate reaches 98% or higher, let the radios run and collect real time statistical data for five to thirty minutes. (Let radio run longer when BER still displays 0)

Step 9: Click Stop radio button on the receiver radio and record radio distance and statistical information from the Statistics Frame

Repeating Steps 1-9 for various separations of the transceivers resulted in the data collected in Table 3-3.

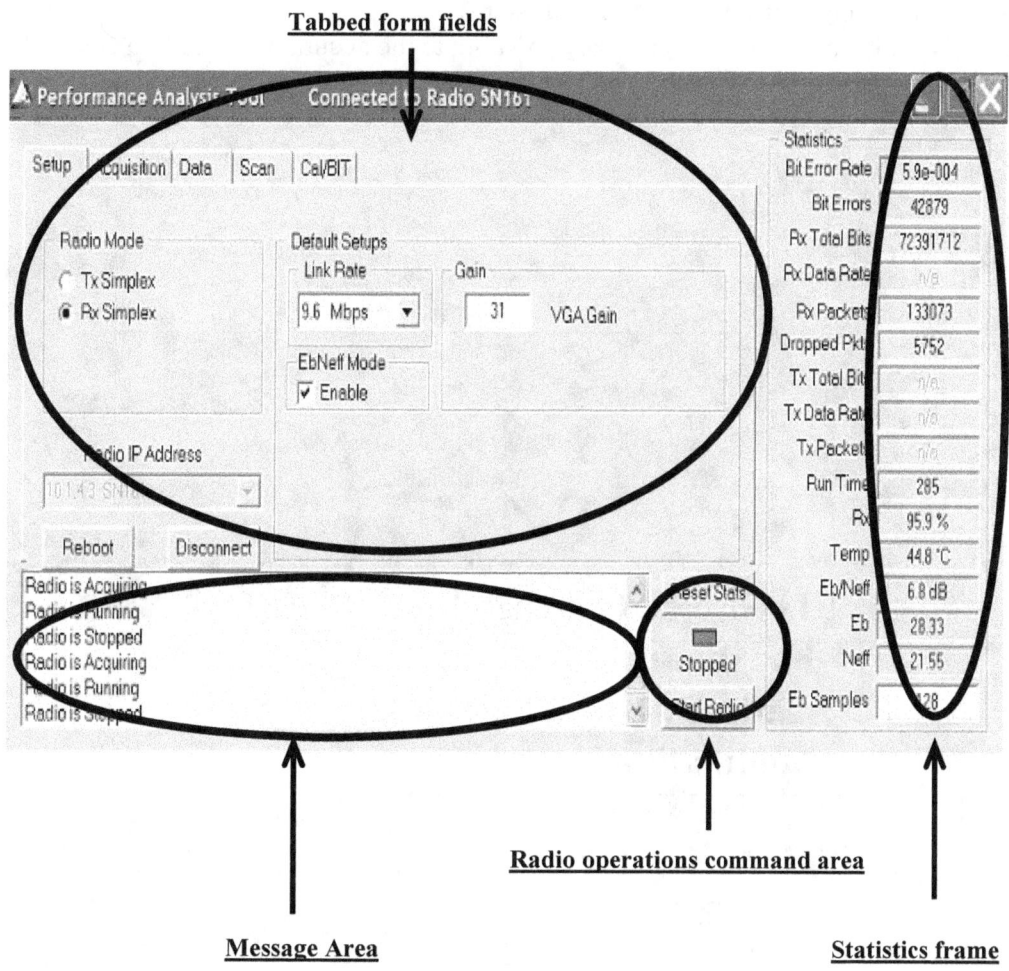

Figure 3-5   Performance Analysis Interface For PulsON 200 Receiver

To validate the PulsON 200 statistical data, the theoretical BER curve for a binary antipodal modulation scheme was calculated and graphed, using the default bi-polar flip modulation while operating the PulsON 200 link (See Figure 3-6).

The theoretical data of Figure (3-6) were graphed from the following equations:

$$BER_{(Flip)} = Q\left(\sqrt{2 \cdot \frac{E_b}{N_{eff}}}\right) \quad (3-8)$$

where

$$Q(x) = \frac{1}{2} \cdot erfc\left(\frac{x}{\sqrt{2}}\right)$$

and where E_b/N_{eff} ranged from 0 to 14, in unit steps.
Development of this curve became necessary to validate the accuracy of the PulsON 200 statistical data.

$$BER_{(Flip)} := Q\left(\sqrt{2 \cdot \frac{E_b}{N_{eff}}}\right)$$

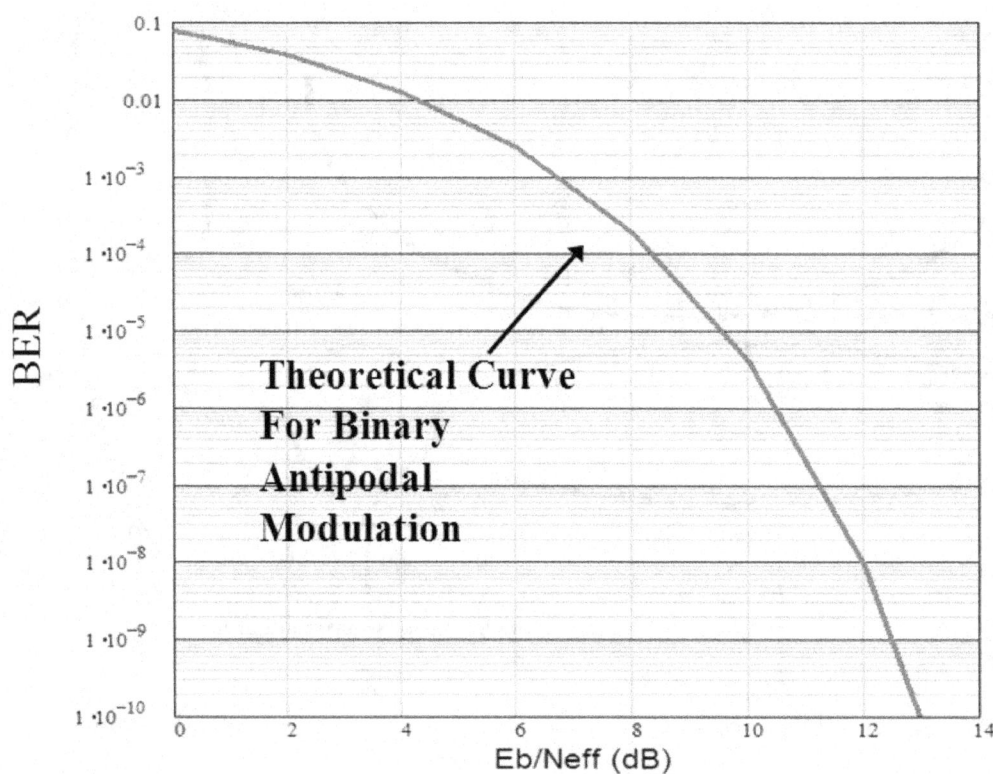

Figure 3-6   Theoretical BER vs Eb/No For PulsON 200 Flip Modulation

In a following section, Figure 3-15 displays the theoretical curve plotted in Figure 3-6 in conjunction with Eb/No vs. BER data points collected from the PulsON 200 test cases collected in Table 3-3. These comparisons confirm that the received data statistics gathered with the PulsON 200 PAT agree well with its theoretical BER curve displayed above, and therefore validates the collected data from the PulsON 200 transceiver for the performance down to where flaring in the BER rate curve occurs.

After validation that the PulsON 200 collected statistical data closely followed the theoretical curve, the measure of the residual BER of the single-banded PulsON 200 link was begun. By measuring the residual BER of the single-banded PulsON 200 link, it was possible to make measurements of a single sub-banded data stream. Then by extrapolation the equivalent timing uncertainties inherited in a single data stream for estimating the total timing uncertainty in a set of data stream were determined. By measuring the residual BER of a single band link, the irreducible timing errors due to a single path were determined. The summary result is that timing errors introduce an effect that limits the probability of bit error such that increasing Eb/No (through increasing transmitter power) does not improve BER below a certain error rate.

3.20 RESULTS OVERVIEW

As was stated previously, the purpose of this UWB research is to determine how accurate timing instances must be to avoid entering the asymptotic region of BER flaring at low BERs in the resultant BER curves. To review, this study examined and predicted the flaring in the bit error curves that occur for the different values of normalized timing jitter variances. To validate the study, empirical comparisons were made using experimental results gathered with a pair of PulsON 200 UWB Evaluation Kit Transceivers and Performance Analysis Tool (PAT). Following are the results from the statistical analysis and the experimental analysis. For the purpose of presentation, this section has been divided into four parts. The first part shows the BER effects due to timing synchronization errors. The second part displays results from the PulsON 200 UWB Radio test cases. In the third part, a relationship is shown with analytical and experimental results. Finally, the fourth part concludes the section with a detailed discussion of how the research results of this paper can be applied to designing systems at all data rates and discusses future research for applying this basic theoretical technique to multi-carrier UWB-OFDM systems.

3.21 BER EFFECTS DUE TO TIMING ERRORS

Bit synchronization information traveling from the transmitter to the receiver must be recovered accurately in digital communication systems. Because in practical digital communication systems we typically transmit only the bit stream and regenerate the bit clock through clock and data recovery (CDR) as required for enabling the taking of the bit samples required for making soft (initial) bit decisions), we introduce bit errors. These bit errors are caused by distortions and noise in the received bit stream along with imperfections in bit clock regeneration (see Figures 3-7 and 3-8). Since the research approach involved reducing bit error rates to as low as possible in UWB systems through identifying maximum timing synchronization errors, we evaluated BER effects due solely to timing jitter, while recognizing that additional effects are often observed in practical hardware implementations.

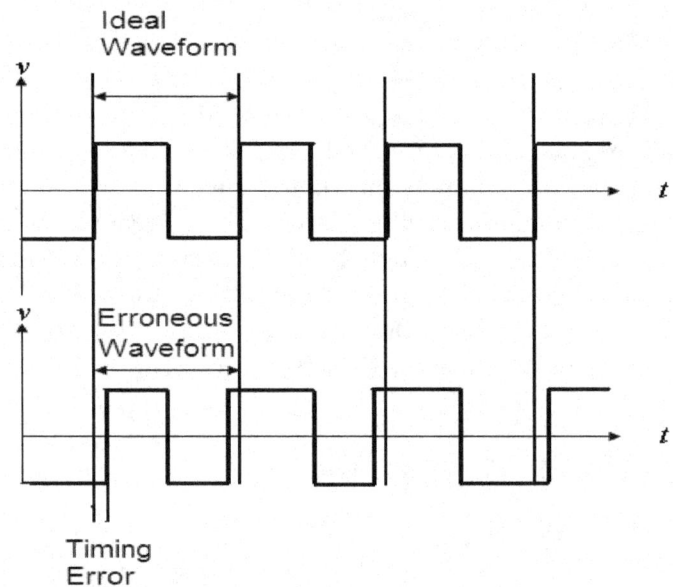

Figure 3-7   **Representation Of Timing Errors In A Digital Signal**

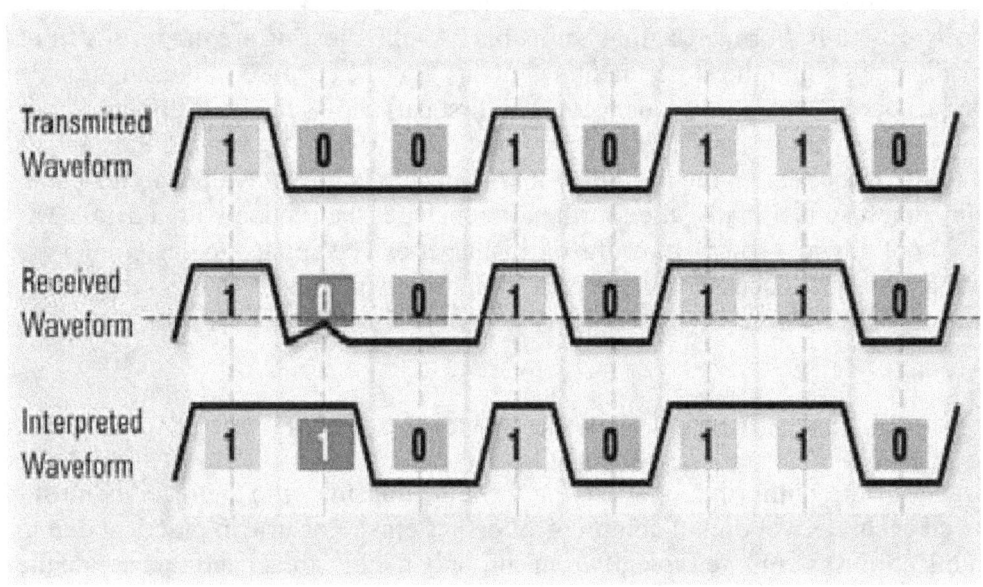

Figure 3-8   **Representation Of Distortions In A Signal**

In the statistical analysis we considered effects of timing jitter on a system's BER performance. The first step in this process was to obtain equations that would derive the error probability of the correlation detector conditioned on a bit synchronization error.

Then we took this conditional error probability and averaged it over the probability density function (pdf) of the synchronization error, yielding the average error probability at the receiver [2]. As a result, we graphically displayed the process mentioned above through Equations (3-1), (3-2), (3-3), (3-4) from the previous section. We plotted these results with MathCAD; the generated plots are shown in Figures 3-9, 3-10, 3-11 and 3-12, respectively.

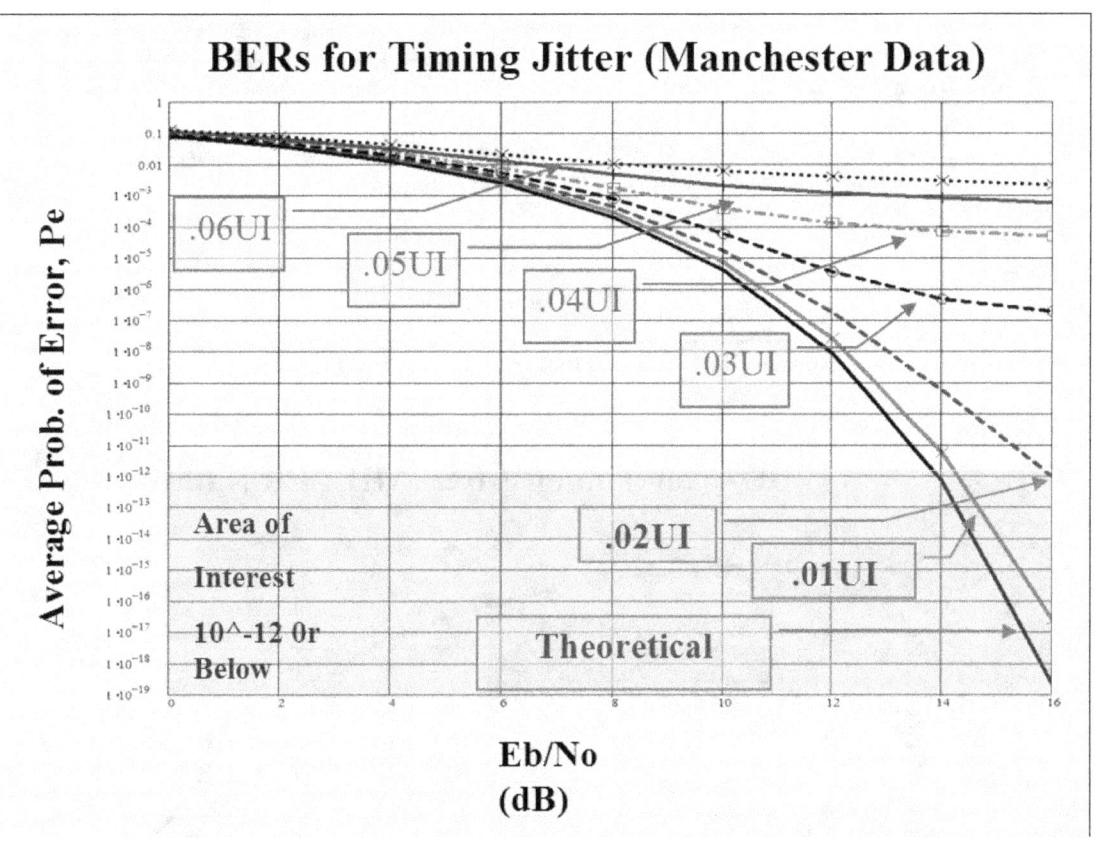

Figure 3-9   **BERs For Timing Jitter (Manchester Data)**

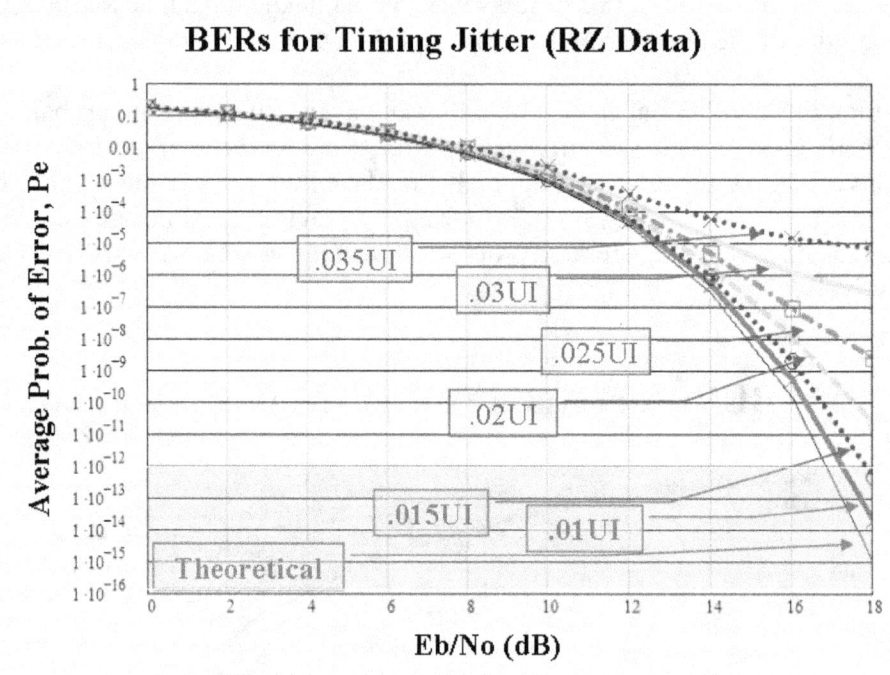

Figure 3-10   BER For Timing Jitter (RZ Data)

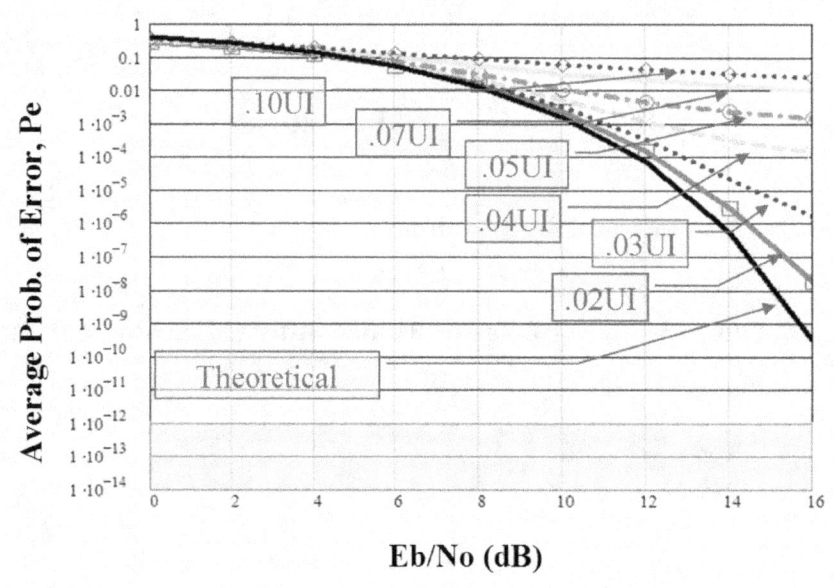

Figure 3-11   BERs For Timing Jitter (MILLER Data)

BERs for Timing Jitter (NRZ Data)

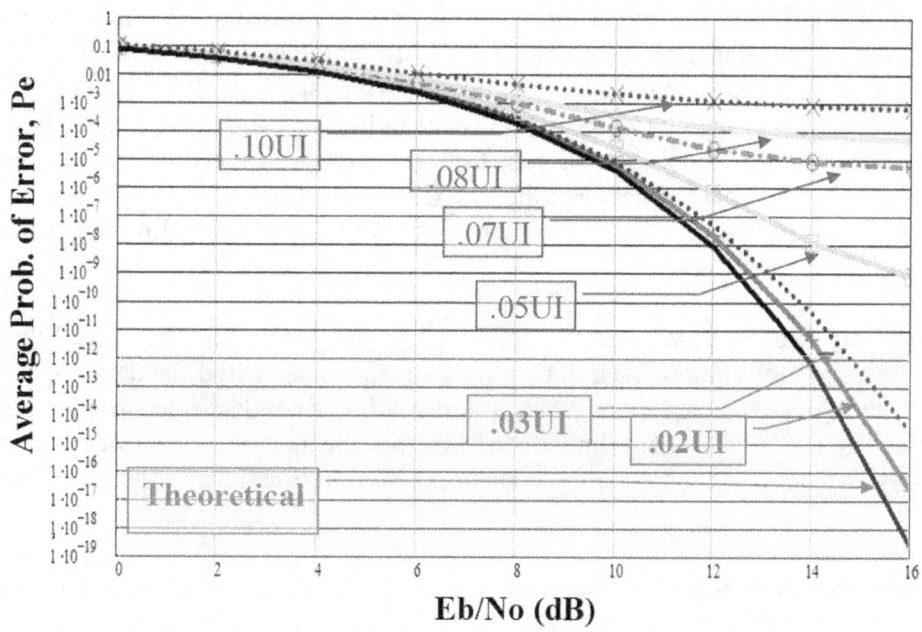

Figure 3-12   BERs For Timing Jitter (NRZ Data)

Analysis of the data from Figures 3-9 through 3-12 determined how accurate normalized timing error variances must be to avoid entering the asymptotic region of BER flaring. Given that the flaring shows the minimum irreducible BER a system with timing errors can achieve at any arbitrarily high transmitter power level, before incorporating forward error correction codes, we estimate how far the bit errors can be pushed down, before the probability of error stops improving (Thereby identifying the flaring points of the UI curves). The Manchester coded graph showed the amount of timing uncertainty that can be budgeted for a transmitter de-multiplexer and the receiver clock and data recovery multiplexing operation. Likewise, NRZ, Miller, and RZ graphs showed the same effect as the Manchester graph, except they may be used for designing systems which incorporate NRZ, Miller, or RZ data in their digital communication system schemes.

To apply the statistical analysis, in the next section we describe a test case that we developed and performed, to approximate or bound the actual timing uncertainty that exists in PulsON 200 radios. As this technique utilizes an easy to implement statistical data collection technique to determine an otherwise difficult-to-determine stochastic jitter performance, it has particular merit whenever making measurements at faster data rates where test equipment performance has not yet caught up to the performance level necessary for directly assessing jitter performance.

3.22 **PULSON200 TEST CASE**

Following the analyses shown earlier, we performed test cases to correlate a statistical analysis with actual hardware analysis. The experimental analysis was achieved with PulsON 200 UWB transceivers and Performance Analysis Tool (PAT) software. PulsON 200 radio technology uses a true UWB pulse, as defined by the FCC, and analysis statistical information about the data sent from transmitter to receiver, is analyzed in real time with PAT. Data passed between the PulsON 200 radios allow an evaluator to configure command and receive performance of UWB data. (See [5] for specific PAT user operations)

For simplicity, the test cases used the default bi-polar FLIP modulation (see Figure 3-13) as the choice of modulation. To conduct the analysis, we selected the data rate to be 9.6 Mbps. Although next-generation UWB systems will deliver data rates in the hundreds of mega bits per second, this experiment confirms that the analysis performed in this paper, showing flaring due to timing errors, is demonstrated in practice by the PulsON 200 radios.

In the following section we outline the equipment list and test setup used to conduct the experiment.

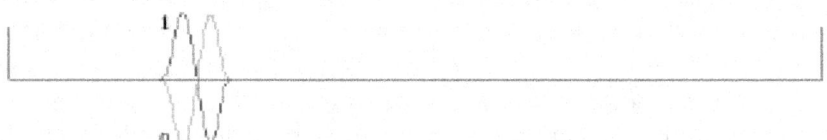

Figure 3-13   **Flip Modulation**[9]

---

[9] Picture from K. K. Lee UWB presentation of Flip Modulation

3.23   REQUIRED EQUIPMENT AND TEST SETUP

To adequately test the UWB radios the following equipment was used:

(2) Time Domain PulsON 200 UWB transceivers
(2) Laptop Computers with PAT software version 3.0 configured and installed
(2) Category 5 Ethernet crossover cables to connect laptops to UWB radios
(1) RS-232 cables to change radio IP address or view calibration test in hyper terminal under Microsoft Windows XP
(1) Measuring tape

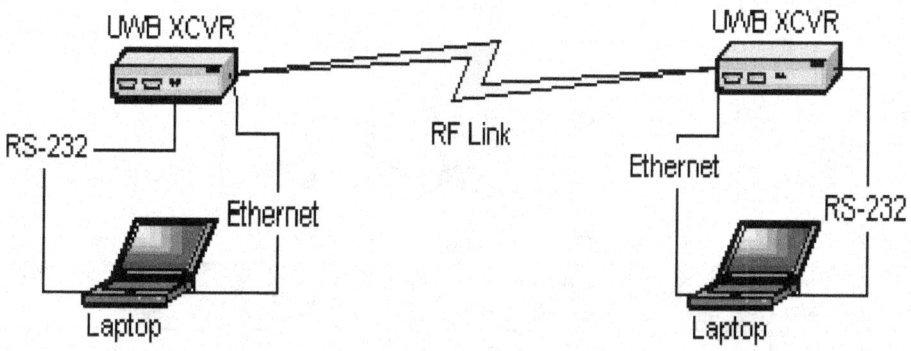

Figure 3-14   PulsON 200 Evaluation Kit Setup[10]

Using the test setup in Figure 3-14, the objective became to acquire an appropriate amount of statistical data to track a theoretical FLIP modulation BER curve and determine indirectly the timing jitter achieved, consistent with the BER flaring performance actually achieved. Results from these test cases are shown in the following section.

3.24   TEST CASE RESULTS

As described previously, statistical analyses about the data transmitted and received from the PulsON 200 transceivers were obtained with the PulsON 200 PAT. To obtain efficient data, we continuously passed data between the two transceivers for time periods ranging from approximately 40 to 1600 second per distance, per test case, as shown in Table 4-1.

---

[10] Picture from [6]

Then, to confirm the experiment result, we examine how close the results were to the well-known theoretical formula for binary antipodal modulation:

$$BER_{(Flip)} = Q\left(\sqrt{2 \cdot \frac{E_b}{N_{eff}}}\right)$$

Notice in Table 3-3 the collected values for BER and Eb/No. Plotting these values, in conjunction with the theoretical curves, shows a close similarity between the curves. This indicates that the collected date confirms the theory. However, a gradually asymptotic BER flare evolves around 10^{-4}. This flaring in the BER curve leads to a discussion, presented later, that allows one to estimate the approximate timing uncertainty inherent with the PulsON 200 transceivers.

Table 3-3 PulsON 200 Collection Of Data Rate Vs Range Test Cases

RANGE vs. DATA RATE 9.6Mb/s									
PulsON 200 Receiver Test Cases									
Statistics	26.24 feet	26.24 feet	26.24 feet	26.24 feet	100 feet	200 feet	200 feet	330 feet	345 feet
BER	2.90E-04	9.40E-04	1.30E-03	1.40E-04	4.70E-04	1.30E-05	7.00E-04	6.90E-03	1.60E-02
Bit Errors	23426	74866	11031	12275	139660	5647	95761	577526	375518
RX Total Bits	80633312	79511040	86669536	88506624	297129536	425477632	136776832	83864128	23567712
Rx Packets	148223	146160	159319	162696	546194	782128	251428	154162	43323
Dropped Pkts	22	19	218	217	122	7	316	4199	38199
Run Time (Seconds)	304	300	327	304	1119	1603	516	325	167
Rx Rate	99.90%	99.90%	99.90%	99.90%	99.90%	99.90%	99.90%	97.30%	53.10%
Temp.	46.3C	45C	47.8	47.5C	44.5C	45.0C	45.0C	45.0C	46.5C
Eb/Neff (dB)	9.7	8.8	13.3	14.8	7.9	13.6	12.6	6.1	4.5
Eb	31	25.52	34.61	34.9	29.01	33.79	32.22	26.37	25.18
Neff	21.29	16.69	21.35	20.13	21.1	20.22	19.63	20.26	20.72
Eb Samples	128	128	128	128	128	128	128	128	128
VGA Gain	31	23	27	31	24	24	24	31	31
Thres. Const.	80	80	80	80	84	82	82	80	83
log (BER)	-3.54	-3.03	-2.89	-3.85	-3.33	-4.89	-3.15	-2.16	-1.80

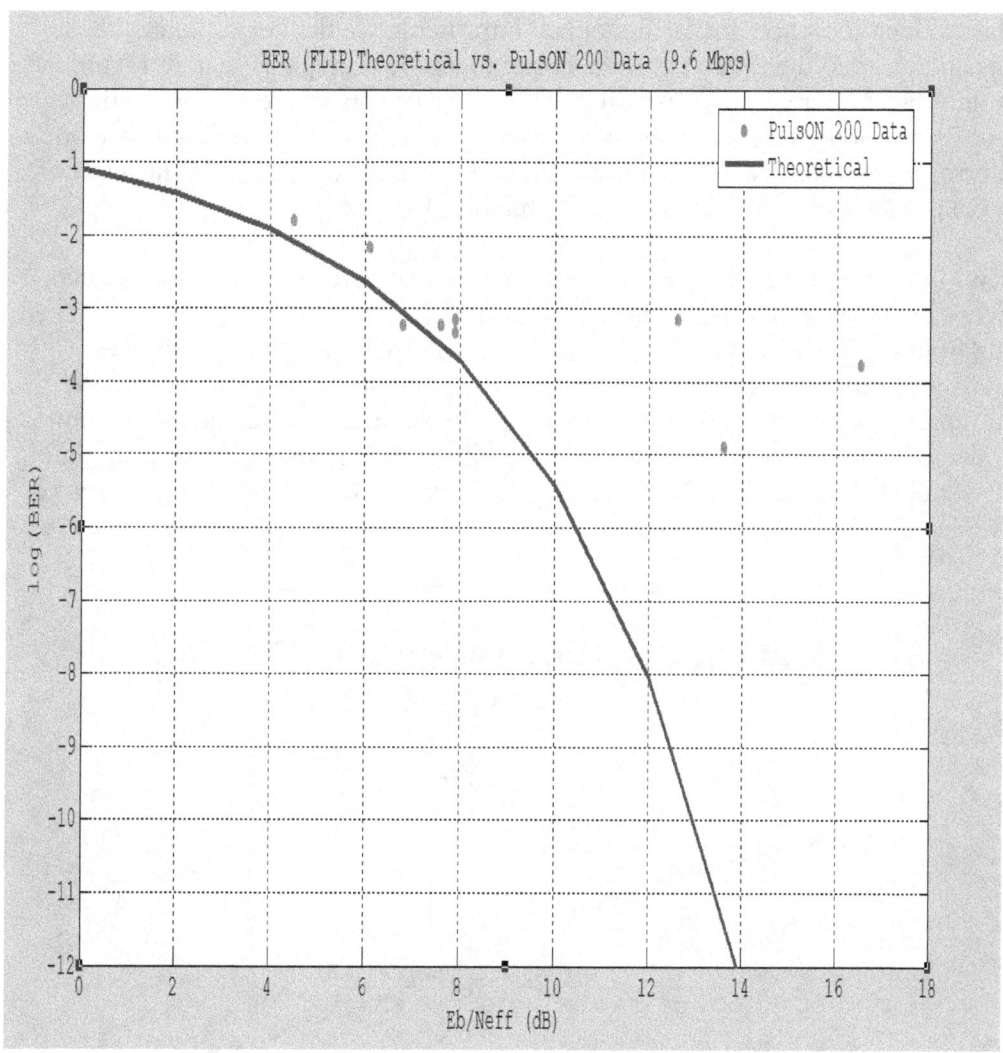

Figure 3-15   Flip Modulation Theoretical Curve vs PulsON 200 Data (9.6 Mbps)

3.25   ASSESSMENT OF TIMING UNCERTAINTIES

For the purpose of comparing Manchester coded data with experimental data collected in Figure 3-15, this section plots the Manchester graph in Figure 4-3 against the experimental data points from Figure 3-15, to assess the approximate residual timing uncertainty inherent within the UWB radio link.

As mentioned previously, the summary result is that timing errors introduce an effect that limits the probability of bit error such that increasing Eb/No does not improve BER below a certain error rate. With the experimental data in Figure 3-15, we found that the probability of error stopped improving in approximately the 10^{-4} region. After collecting enough data to identify where the BER stopped improving, we then extrapolated these data points and inserted them into the Manchester theoretical graph as seen in Figure 3-16. Since the Manchester graph shows probability of error conditioned on normalized timing synchronization error, we were able to identify the normalized unit interval value of timing error in the PulsON 200 radio. We identified this region on the Manchester curve, and expanded timing sync values in the region of 10^{-3} to 10^{-5}.

Graphed in Figure 3-16., the black solid curve represents theoretical values, while the other curves range from normalized variances of 0.055 UI to 0.030 UI. The extrapolated data points from the PulsON 200 experiment is also represented on the graph by the diamond-shaped points. As seen in Figure 3-16., experimental values surrounded the curve that represents 0.04 UI. From this analysis, one can identify the normalized unit interval value of timing error in the PulsON 200 radio. The hypothesis concerning timing uncertainty budgeted for the PulsON 200 radios is identified to be 0.04 UI, since this UI jitter curve best fits the experimental data points.

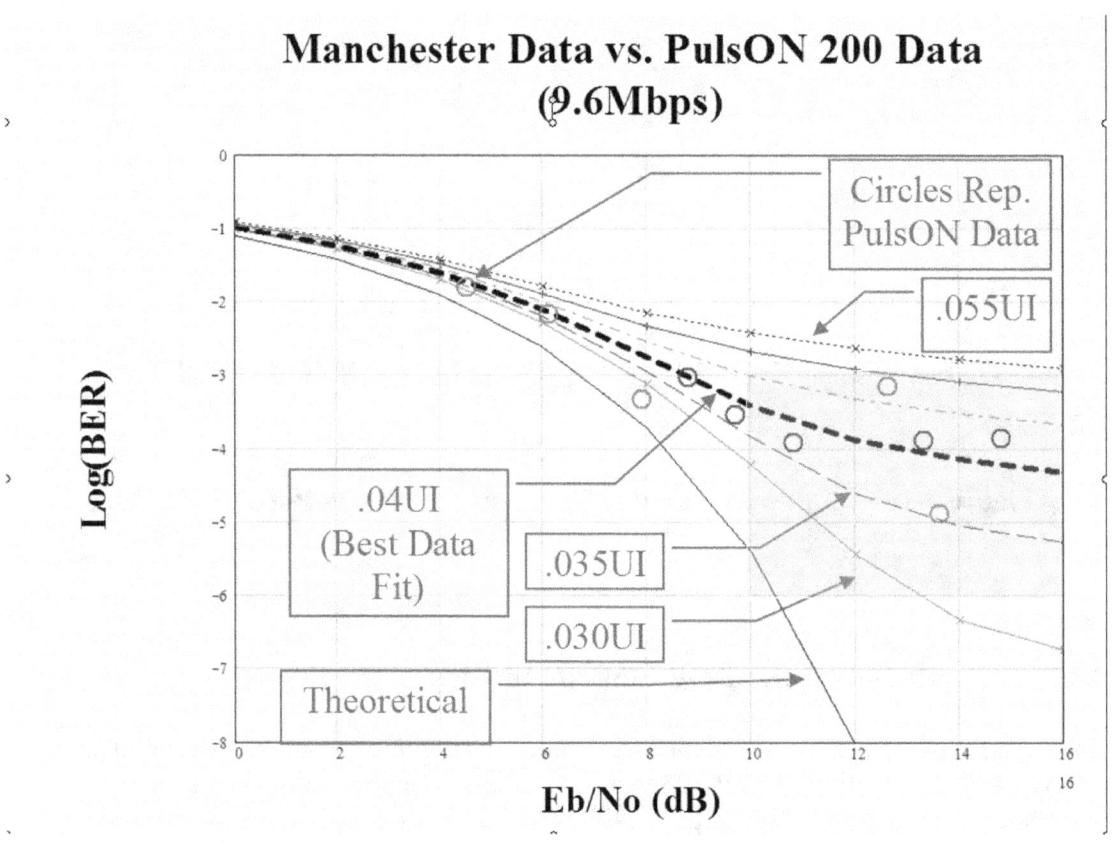

Figure 3-16   Manchester Data With Extrapolated PulsON 200 Data

In conclusion, this research estimated the normalized timing uncertainty for the PulsON 200 radio, due to timing jitter, through an easily measured indirect technique, instead of a difficult-to-make direct technique. One is able to use this same concept to predict how much timing uncertainty should be budgeted for designing future high data rate (hundreds of Mbps) UWB systems. This analysis technique allows one to budget how much timing uncertainty can be allowed while still achieving acceptable margins for the transmitter de-multiplexer and the receiver clock and data recovery multiplexing operations, which leads to the discussions in the next section.

3.26 ANALYSIS OF EFFECTS ON UWB-OFDM SYSTEM WITH HIGH DATA RATES

Predicting how much timing uncertainty that should be used in designing future high data rate systems in the multiple hundreds of mega bits per second can be estimated with the previously identified unit interval (UI) matching approach. A UI is defined as one nominal bit period for a given signaling speed. Therefore, rather than using an absolute timing approach, the UI approach method of analysis can be used in budgeting timing uncertainties for UWB-OFDM systems operating at higher data rates.

Likewise, normalized (with respect to a bit time) timing error UI's are represented as the timing synchronization errors analyzed throughout this paper. These values are small percentages of a unit interval, which define the standard deviation of a normalized timing error in terms of UI random timing jitter. A 0.02 UI random timing error measurement informs that the standard deviation statistic of a bit period deviation is 2% around the ideal bit period time, through utilizing a Unit Interval approach, in place of an absolute timing approach, the results of this paper can easily be applied to ever-increasing data rates of future UWB-OFDM data links considered for meeting NASA's communication needs.

The results in this paper pertain to a reference UWB-OFDM system. In practice, approaching UWB- OFDM systems will likely use a multi-band approach where a cyclic prefix is appended to the beginning of the OFDM symbol. Once the bit timing synchronizations are corrected using the approach discussed in this paper, further research can be applied to analyze the timing symbol synchronization errors in the multi-band UWB-OFDM. In the same way as for the reference UWB OFDM system affected by bit timing errors, the average probability of error may be obtained for the symbol timing errors in the practical UWB-OFDM system. Assessment of timing uncertainty of realistic symbol timing in the practical multi-band UWB-OFDM may be realized through the same approach used to assess the bit time synchronization timing uncertainties in the reference UWB-OFDM system. Qualitatively, the main results from this paper remain valid for the practical multi-band UWB-OFDM systems.

3.27   SUMMARY OF RESULTS

This paper has developed theoretical equations for estimating BER effects due to timing uncertainties among multiple OFDM channels. Having been considered are UWB-OFDM systems which are coded by Manchester, Miller, RZ, and NRZ data. The focus is on reducing BER to as low as possible through identifying maximum timing synchronization errors for each data format. Since pushing bit errors to below 10^{-12} before using forward error correction codes is advocated, the maximum reserve can be maintained for the FEC hardware to correct for errors caused by instances other than timing errors. Specifically, this paper identifies how accurate timing error instance must be to avoid introducing the asymptotic region of BER flaring at probability of errors of 10^{-12} or below. According to the presented analysis for Manchester data formats, to obtain a Pe of 10^{-12} or below, timing accurate instances should be budgeted at a 0.02UI maximum. NRZ coded data timing accuracy should be budgeted at a maximum of 0.04UI. RZ data formats should be budgeted at a maximum of 0.015UI. Miller Coded data was not able to achieve a probability of error of 10^{-12} for reasonable values of Eb/No. Therefore, it was found that this data format should not be used. These budgetary values provide an estimate of the timing accuracies required for a given BER performance to MUX a set of parallel transmitted, De-Muxed data streams sent in parallel over multiple OFDM data streams, utilizing multiple OFDM symbols transmitted within multiple sub-bands.

The PulsON 200 "EVK" was used to make measurements of a single sub-banded data stream, to determine the equivalent timing uncertainties inherited in a single data stream for estimating the total timing uncertainty in a set of OFDM data streams. With the test cases performed using PulsON 200 we were able to use the analytical solutions to determine the approximate amount of timing uncertainty budgeted for the radios. This test case demonstrated that the analytical and experimental solutions agree well. As a result, we are able to assess how much timing uncertainty that can be budgeted for UWB systems and all other high data rate systems through the unit interval approach.

3.28   RECOMMENDATIONS FOR FUTURE RESEARCH

Recommendations for continuing this research include determining the probability of error effects due to second order degradations other than timing errors. These second order degradations may include probability of error effects due to phase noise, multi-path propagation effects, noise figure.

In addition, looking at timing error effect along with the second order degradations mentions above should be examined for multi-banded UWB-OFDM. Instead of examining the bit synchronization of the data stream as examined in this paper, the idea can be expanded to examine the symbol synchronization effects. This would involve not only timing error effect, but also frequency error effects of the multiple OFDM symbols and the requirement to multiplex the symbols back together in ways that minimize inter-symbol interference and inter-carrier interference, while lowering bit error rates down below 10^{-12}.

ECT - Phase 3

4.0 ECT SUMMARY RECOMMENDATIONS FOR CONTINUED RESEARCH

The following major task areas are recommended for continued research in the next fiscal year for emerging communication technology development:

- Networked FSO
- RFID
- Expanded Range Wi-Fi
- UWB

1.) Survey: Conduct an industry survey on networking hardware for implementing a small networked optical FSO system, and procure a fixed, non-tracking, wide-beam optical FSO system set of components with redundant optical beams to investigate the limits of networking multiple FSO systems.

2.) FSO Testing:
a. Update previously generated test procedures, adapting and expanding these procedures to account for the multiple FSO OTUs, to permit testing the FSO system exemplars for Bit Error Rate, and throughput rates versus weather-induced degradations (e.g., fog, rain, etc.) when operating in a networked configuration.

b. Test the performance limits of this FSO hardware within the unique environment of KSC, with data path links over both water and over land, comparing the applicability of this technology to KSC's needs versus the single FSO link tested previously

3.) UWB Analysis:
a. Analyze new, OFDM multi-band UWB hardware for applicability on the Range, with particular emphasis on piconet sub-division capability within LANs, recurring costs of hardware, and life-cycle operational costs.

b. Analyze position-aware capabilities of UWB communication technology, with particular emphasis on determining positional accuracy limitations (e.g., accuracy in centimeters, ability to provide relative and absolute positional information.)

4.) UWB Testing:

ECT - Phase 3

a. Update previously generated test procedures, adapting and expanding these procedures for testing the UWB evaluation kit for position aware functionality.

b. Test position-aware capabilities of UWB communication technology using the UWB evaluation kit (EVK) procured during FY03, following the position aware enhanced test procedures developed from previously generated test procedures.

c. Review current UWB and FSO products and theoretical developments through attending two major optical communication conferences and one joint NASA-USAF Advanced Range and Spaceport Technology Conference.

These activities are needed to achieve the 24/7, always-on, highly-mobile vision of an interconnected communication for use on the Range employing First Mile / Last Mile extensions to the existing Range communication infrastructure.

APPENDIX A:
UWB EVK Performance Analysis Tool (PAT) Software

A.1 PAT Statistics Frame Area

The Statistics Frame Area shown on the right side of the Performance Analysis Tool (PAT) shows key PulsON 200 radio performance measurements (See Figure A-1) While observing performance parameters of the connected radio, one is able to analyze: BER, Bit errors, Receiver Total Bits, Receiver Data Rate, Receiver Packets, Dropped Packets, Transmitter Total Bits, Transmitter Data Rate, Transmitter Packets, Run Time, Receiver percentage Rate, Radio Temperature, Energy Per Bit/ Effective Noise (dB), Energy Per Bit, Effective Noise, and Number of Samples Over which Eb is computed.

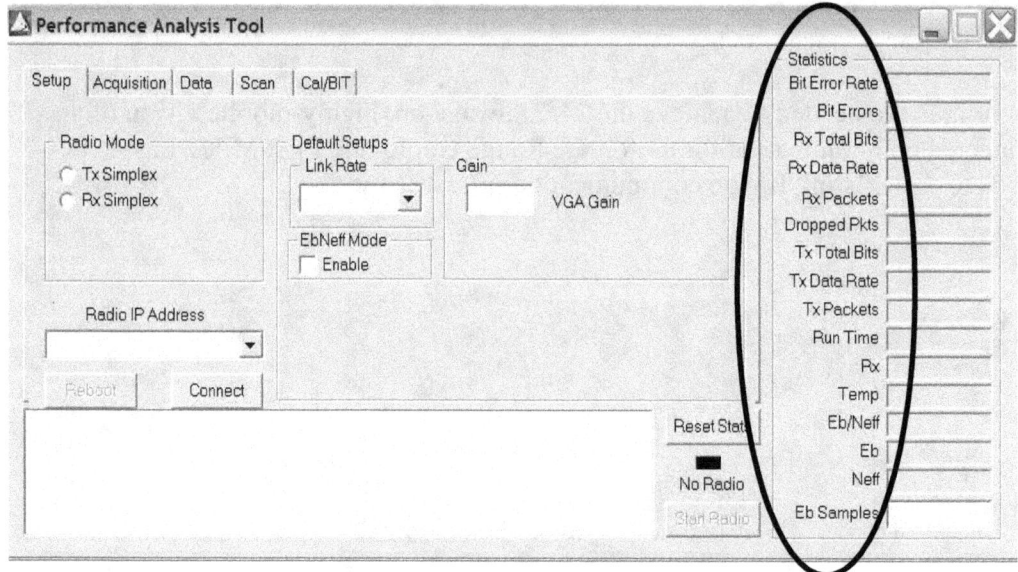

Figure A-1: PAT Statistics Frame Area

The above statistical parameters were defined and computed automatically with the PAT by the following (As outlined in [5]):

Bit Error Rate (**BER**)- The ration between the number of bits in error and the total bits received, computed as: BER= Bit Errors/ Rx Total Bits (BER)

Bit Errors-Total number of bit errors detected by comparing the received bit pattern with the known transmitted bit pattern.

RX Total Bits- Total number of payload bits received. This number does not include the overhead of the acquisition preamble or the packet header.

RX Data Rate- Rate at which data is being received:
Rx Data Rate = Rx Total Bits/ Run Time (Effective data rate)

RX Packets- Total number of packets received

Dropped Pkts- Total number of packets whose number is not sequential to the packet last received, computed as:
Dropped Packets= Dropped Packets + (current packet number – last packet number-1)

TX Total Bits- Rate at which data is being transmitted, computed as:
Total number of payload bits transmitted.

TX Data Rate- Rate at which data is being transmitted, computed as:
(# of packets received/# of packets sent) x 100

Temp- temperature of the temperature sensor on the PulsON 200 Development Module

Eb-Energy per bit, computed as:

$$E_b := 10 \log\left(\frac{m^2}{N}\right)$$

where

$$m := \frac{1}{R} \cdot \sum_{i=1}^{R} (r_i)$$

N= 1 for Flip modulation, 2 for QFTM, 4 for QFTM4
r= raw positive ramp value with calibrated DC offset applied
R= number for ramp

Neff- Effective noise computed as:

$$N_{eff} := 10 \log(2 \sigma^2)$$

$$m := \frac{1}{R} \cdot \sum_{i=1}^{R} (r_i - m)^2$$

R= number for ramp
r =raw positive ramp value with calibrated DC offset applied (when running a normal link)
OR
=raw positive or negative ramp with calibrated DC offset applied (when capturing Ambient RF)

Eb/Neff- Energy per bit/ Effective Noise, computed as:
Eb – Neff (dB)

Eb Samples- This value is the number of samples over which Eb is computed. The default number of samples is 512 and must follow the rule: 2<= Eb Samples <= 4095.

A.2 PAT Range vs. Data Rate Test Cases

The Range vs. Data Rate Test used in this research tested throughput rates over various distances between the UWB transceivers. Test Cases performed during the testing proceedings are shown below through screen shoots taken from the Performance Analysis Tool (PAT).

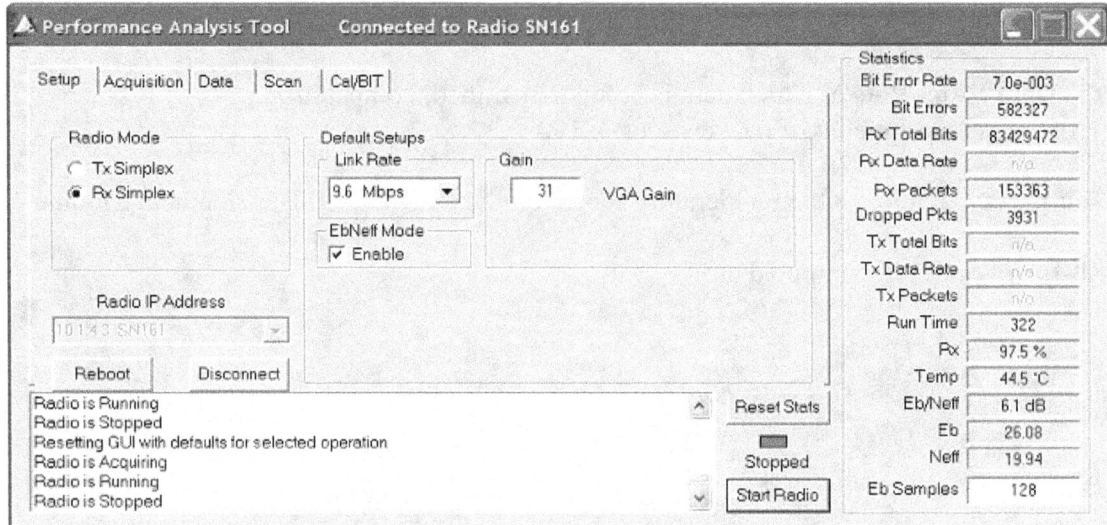

Figure A-2: PAT Test Case @ 330 feet distance

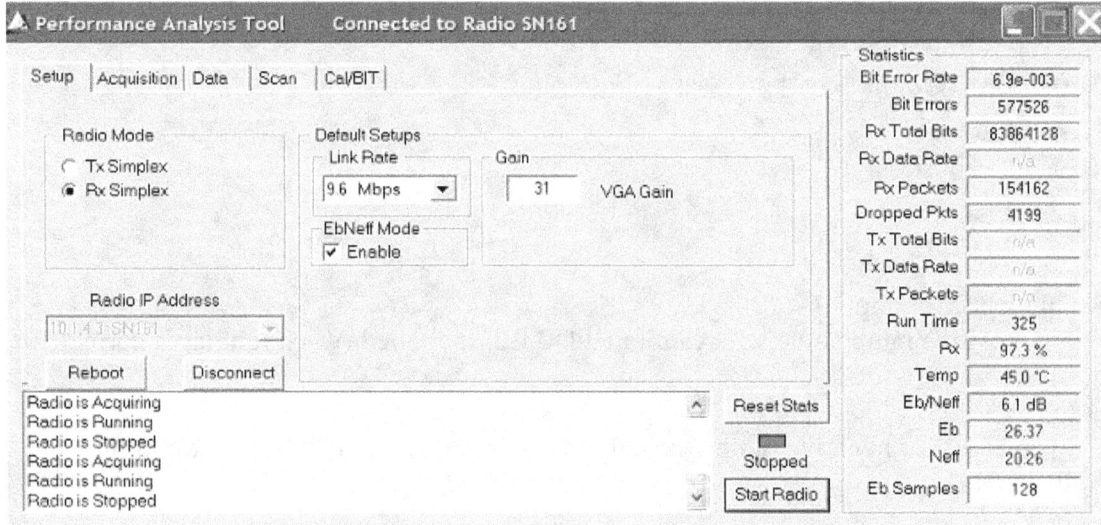

Figure A-3: PAT Test Case @ 330 feet distance

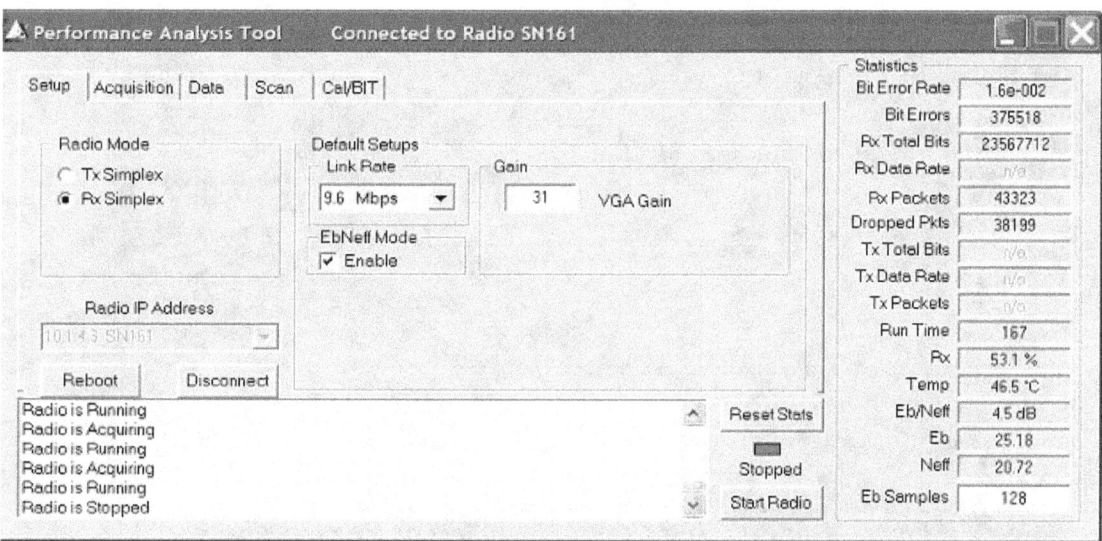

Figure A-4: PAT Test Case @ 345 feet distance

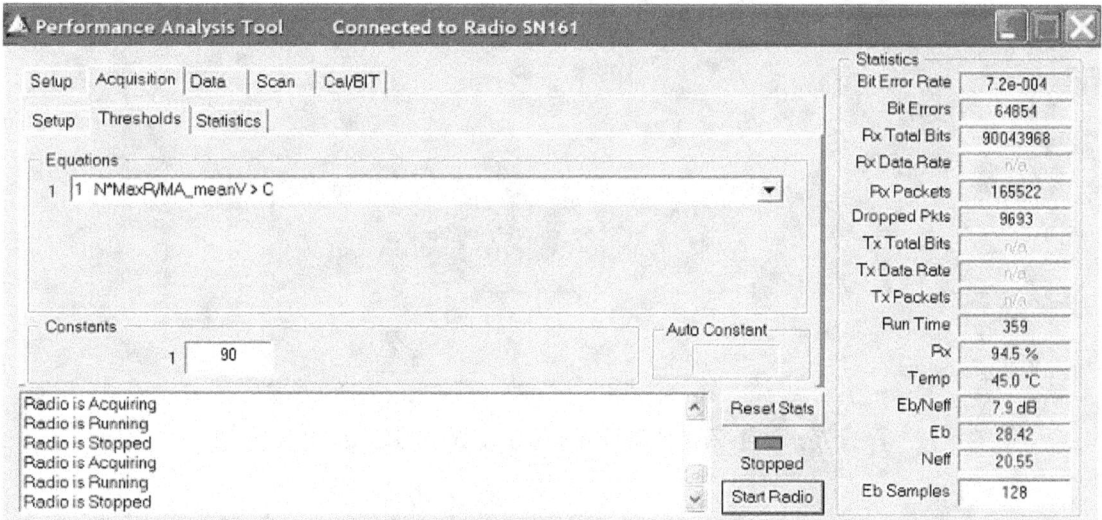

Figure A-5: PAT Test Case @ 300 feet distance and VGA 31

By extracting the Eb/Neff value and its corresponding BER from each test case above, we were able to plot and analyze the data and make hypotheses throughout this paper.

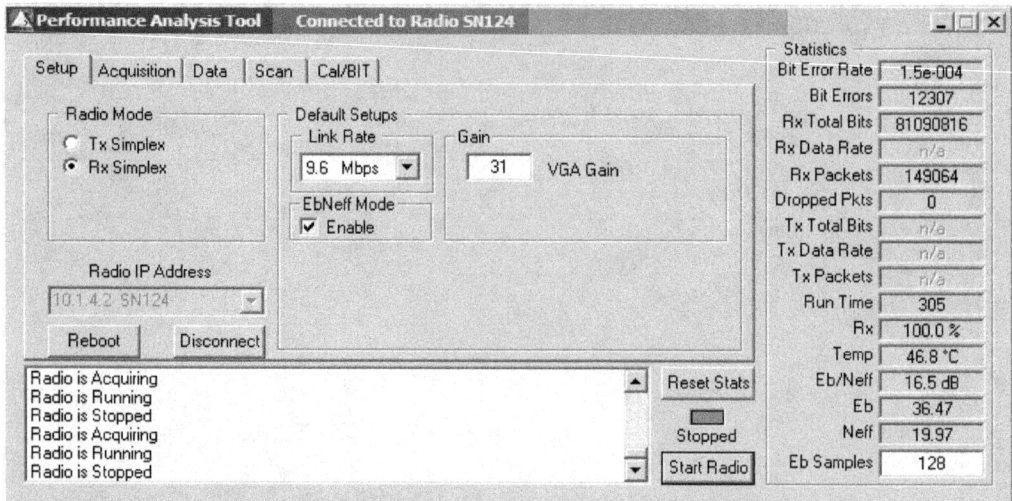

Figure A-6: PAT Test Case @ 16.40 feet or 5 meters distance

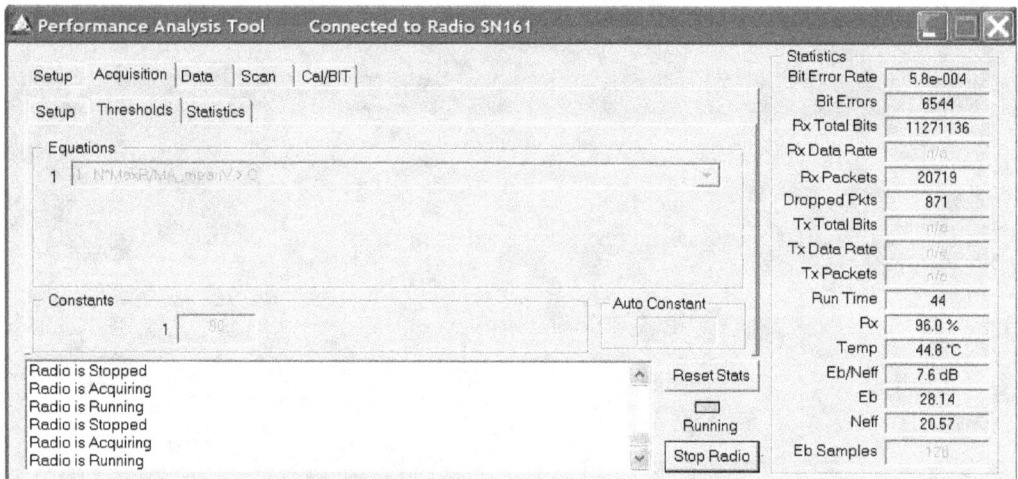

Figure A-7: PAT Test Case @ 300 feet distance

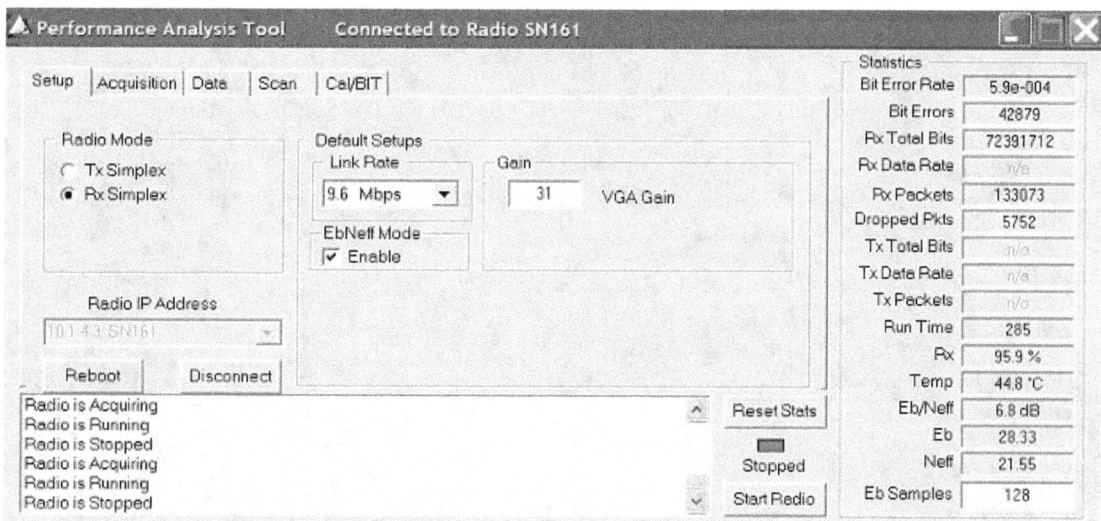

Figure A-8: PAT Test Case @ 300 feet distance

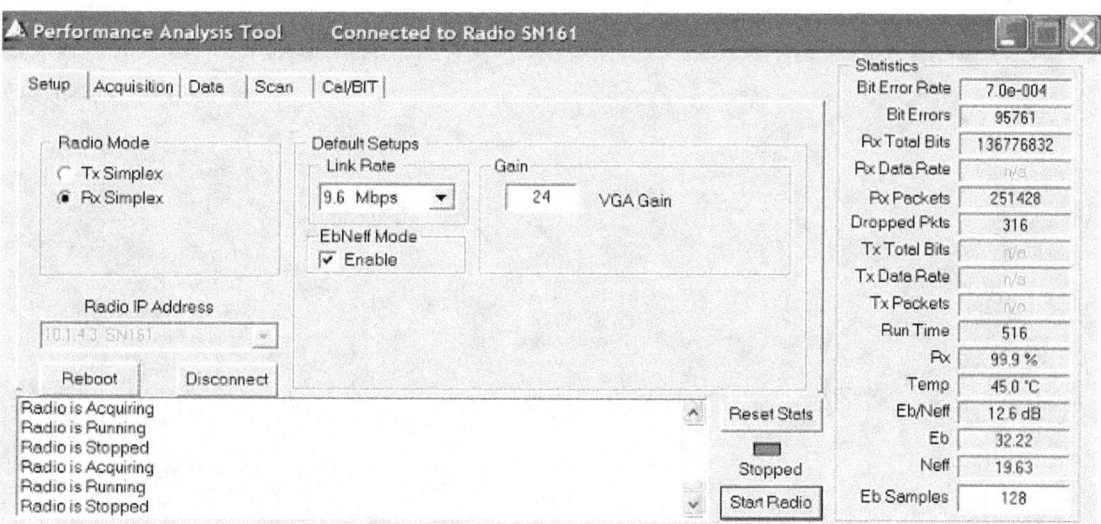

Figure A-9: PAT Test Case @ 200 feet distance

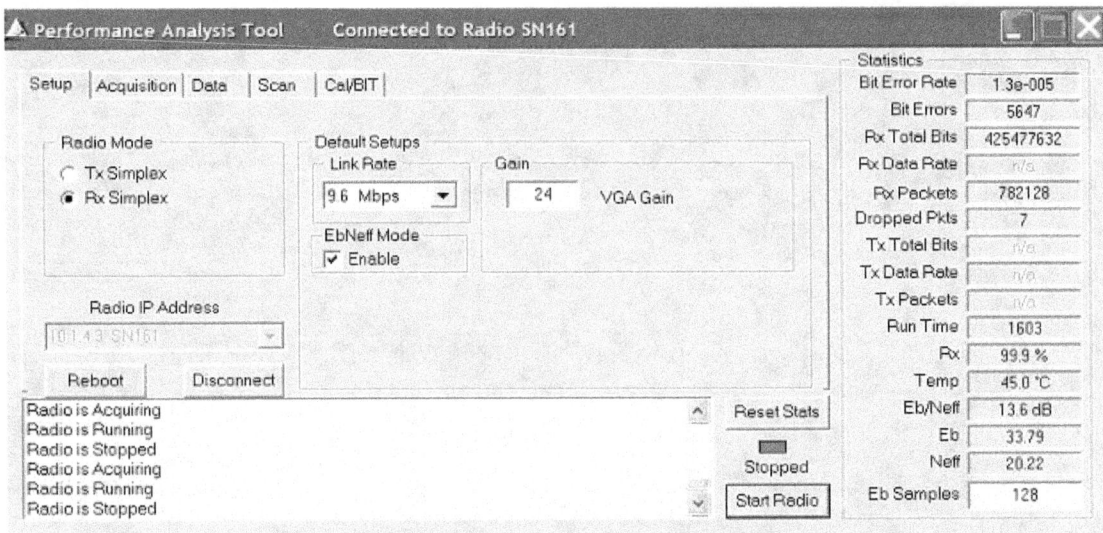

Figure A-10: PAT Test Case @ 200 feet distance

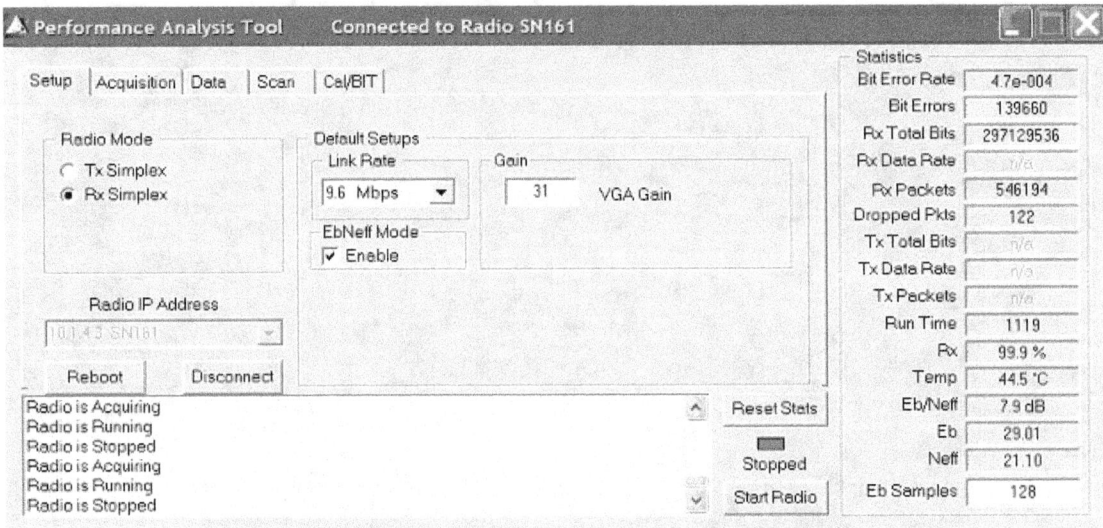

Figure A-11: PAT Test Case @ 100 feet distance

ECT - Phase 3

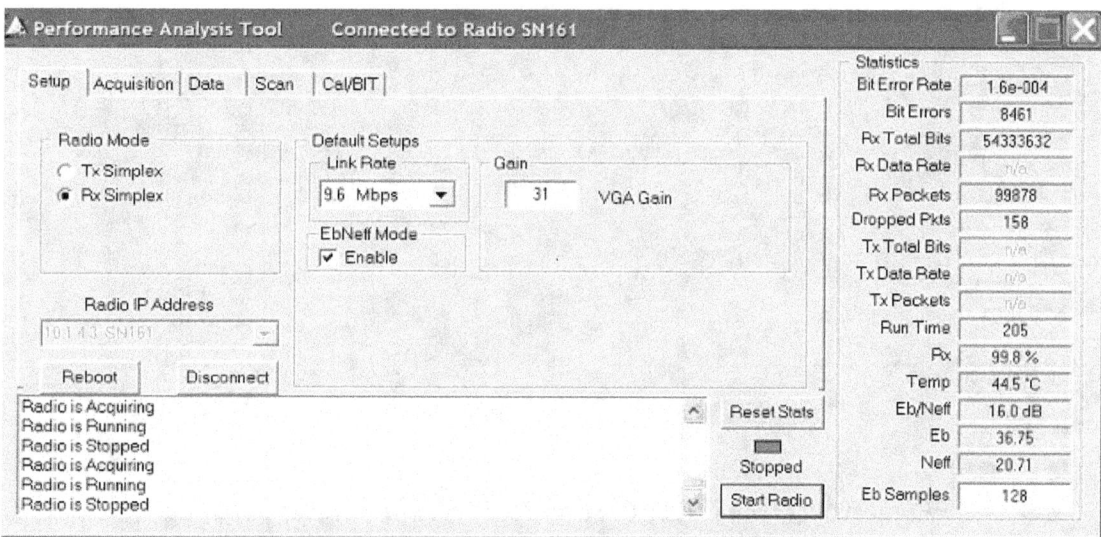

Figure A-12: PAT Test Case @ 10 feet distance

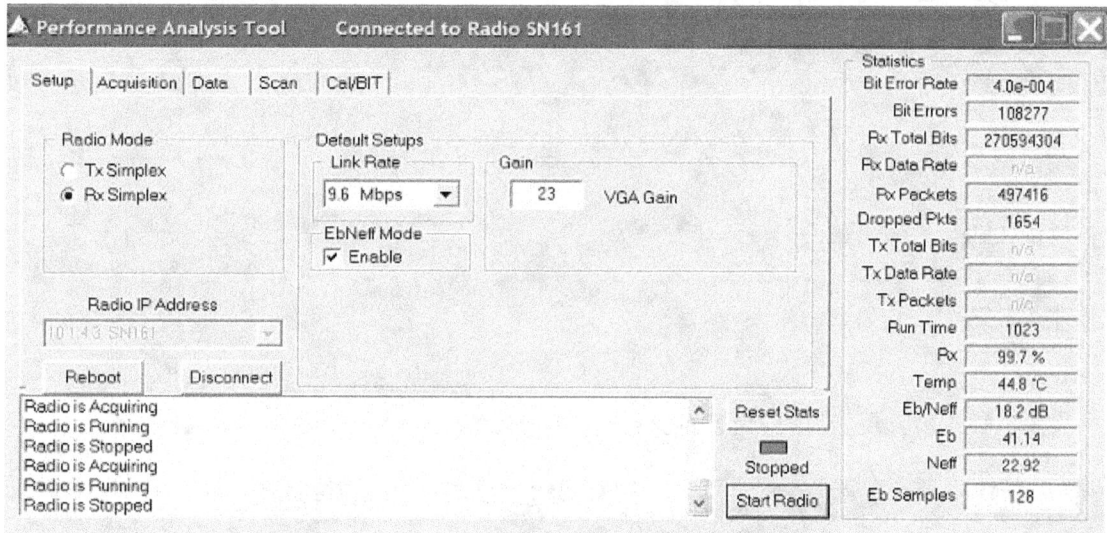

Figure A-13: PAT Test Case @ 5 feet distance

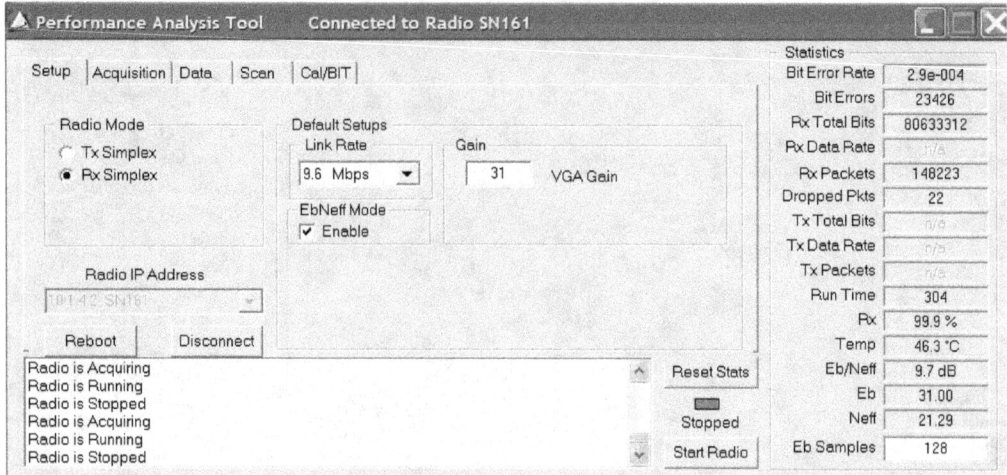

Figure A-14: PAT Test Case @ 26.24 feet or 8 meter distance

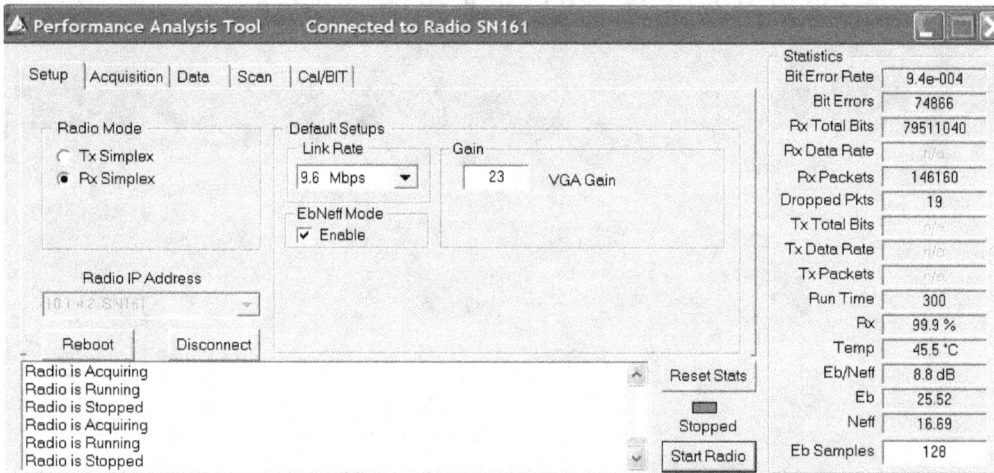

Figure A-15: PAT Test Case @ 26.24 feet or 8 meter distance

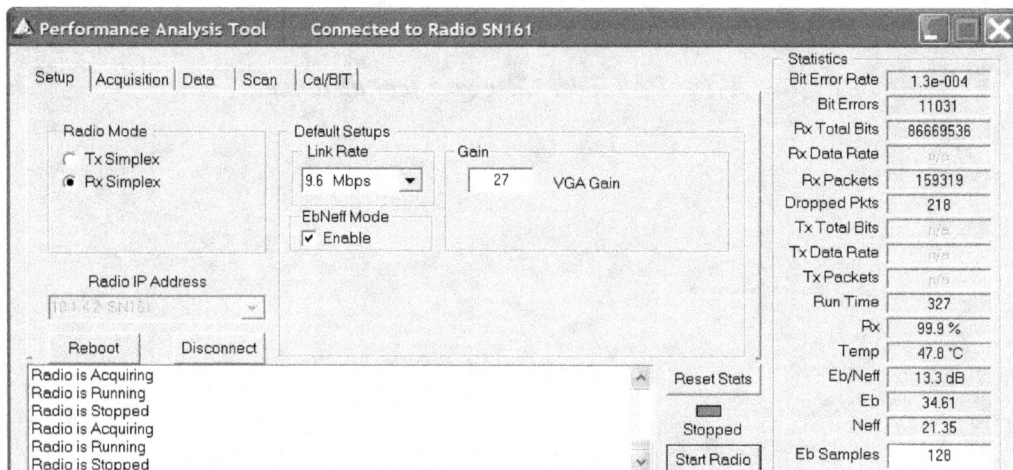

Figure A-16: PAT Test Case @ 26.24 feet or 8 meter distance

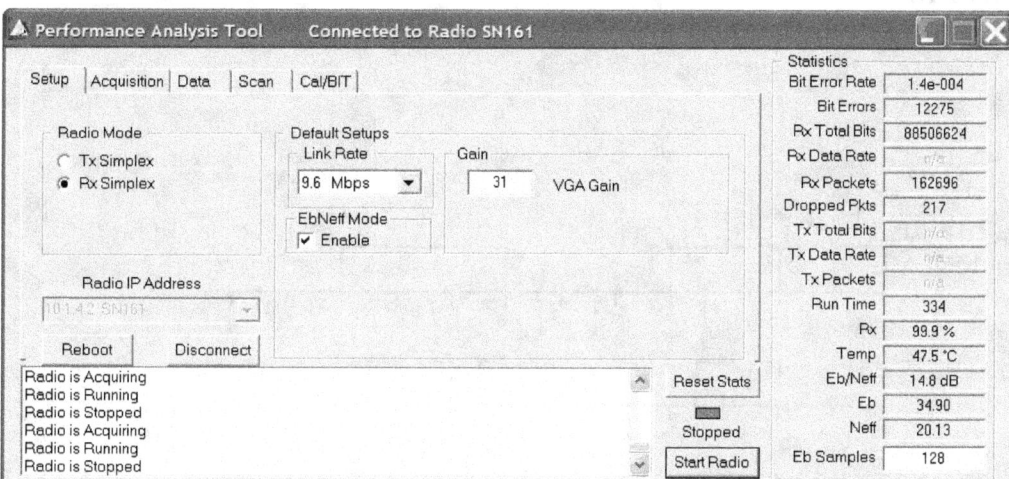
Figure A-17: PAT Test Case @ 26.24 feet or 8 meter distance

APPENDIX B:
Digital Signaling Formats Applicable T0 FSO And UWB Communication Signaling

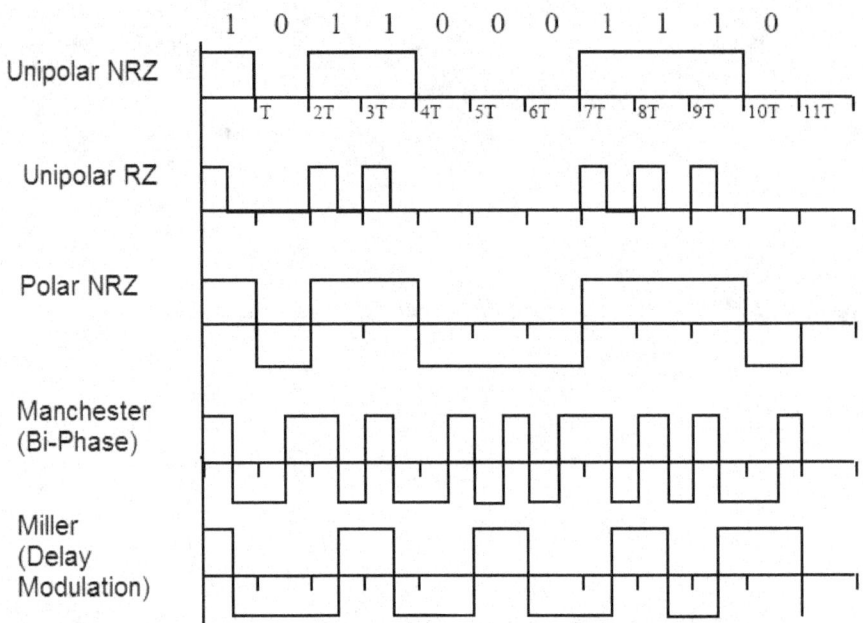

Figure B-1:Binary Line Coding

Binary 1's and 0's may be represented by various binary line codes. Some of the more popular formats are shown in Figure B-1 and are defined as follows.

Unipolar NonReturn to Zero (NRZ) Signaling is a positive logic unipolar signaling where the binary 1 is represented by a high level and a binary 0 by a zero level. This type of signaling is also called on-off keying and is of the NRZ type since the high level does not return to zero during the binary 1 signaling intervals.

Unipolar Return to Zero (RZ) is a unipolar waveform in which a binary 1 is represented by a high level over half of a bit period and then returns-to-zero. The binary 0 is represented by a zero level.

Polar NRZ is binary 1's and 0's that are represented by equal positive and negative levels. This type of waveform is also said to be of the NRZ type.

Manchester coding is where each binary 1 is represented by a positive half-bit period pulse followed by a negative half-bit period pulse followed by a positive half-bit period pulse. This is called split-phase encoding.

Miller line code is where a binary **1** is represented by a transition at the mid-bit position, and a binary **0** is represented by no transition at the mid-bit position. If a **0** is followed by another **0,** however, the signal transition also occurs at the end of the bit interval, that is, between the two **0**s.

APPENDIX C:
Q-Function, ERF, AND ERFC, Applicable To Bit Error Rate Theoretical Analysis Of FSO AND UWB Communication Systems

Q-function is described as the tail integral of a unit Gaussian probability density function (pdf). Represented mathematically:

$$Q(x) = \frac{1}{\sqrt{2\pi}} \cdot \int_x^\infty e^{\frac{-t^2}{2}} dt$$

Other functions that are closely related to *Q(x)* include error function *erf* and complimentary error function *erfc*.

$$\text{erf}(x) = \frac{2}{\sqrt{\pi}} \cdot \int_0^x e^{-t^2} dt = 1 - \text{erfc}(x)$$

[23]

and

$$\text{erfc}(x) = \frac{2}{\sqrt{\pi}} \int_0^\infty e^{-t^2} dt = 1 - \text{erf}(x)$$

[23]

Relationships between *Q(x)*, *erf*, and *erfc* can be shown by the following:

$$Q(x) = \frac{1}{2} \cdot \left(1 - \text{erf}\left(\frac{x}{\sqrt{2}}\right)\right) = \frac{1}{2} \cdot \text{erfc}\left(\frac{x}{\sqrt{2}}\right)$$

In scientific literature there are a few variations of *erfc* which differ by definition. For example in literature reference by Harry Van Trees, *erfc* is defined differently than the classical mathematical definition in material referenced by Abramowitz and Stegun. The choice made in this paper is to use the classical definition, which is more commonly used in the published literature

REFERENCES

1. Harada, Hiroshi, Prasad, Ramjee, Simulation and Software Radio for Mobile Communications, Artech House, 2002.
2. Lindsey, W. C. and Simon, M. K., Telecommunication Systems Engineering, Prentice-Hall, Inc., 1973.
3. Athaudage, Chandranath, "BER Sensitivity of OFDM Systems to Time Synchronization Error," Proceedings of the IEEE, vol.1 (2002): 42-46.
4. Heiskala, Juha and Terry, John PhD, OFDM Wireless LANs: A theoretical and Practical Guide, SAMS Publishing, 2002.
5. "PulsON 200 Evaluation Kit Users' Manual," Time Domain Corporation, 2001-2002.
6. Bastin, Gary L.; Harris, William G.; Chiodini, Robert; Nelson, Richard A; Huang, Po Tien; and Kruhm, David A., "Emerging Communication Technologies Phase 2 Report," NASA/TM-2004-211522, vol. 3, September 2003.
7. Couch, Leon W. II, Digital and Analog Communication Systems, Second Edition, Macmillan, Inc., 1987.
8. Ziemer and Tranter, Principal of Communication, Fifth Edition, John Wiley and Son, 2002.
9. Aiello, G. Roberto and Roberson, Gerald D., "Ultra-Wideband Wireless Systems," Proceedings of the IEEE microwave magazine, June 2003.
10. Win, Moe Z., Scholtz, Robert A, "Impulse Radio: How It Works," Proceedings of the IEEE, vol.2 (1998): 36-38.
11. Retrieved July 28, 2004 from, http://www.fcc/gov/Bureaus/Engineering_Technology/Orders/2002/fcc02048.pdf.
12. Ramirez-Mireles, Fernando, Scholtz, Robert, "System Performance Analysis of Impulse Radio Modulation", Proceedings of IEEE, RAWCON Conference, August (1998), pgs: 67-70.
13. Retrieved August 2, 2004 from, http://www.intel.com/technology/ultrawideband/downloads/Ultra-Wideband_Technology.pdf.
14. Retrieved June 2, 2004 from, http://www.win.tue.nl/~hmei/Personal/MSc_thesis_HailiangMei.pdf
15. Van Nee, Richard and Prasad, Ramjee, OFDM for Wireless Multimedia Communications, Artech House Publisher, 2000.
16. Bingham, John A. C., "Multicarrier Modulation for Data Transmission: An Idea Whose Time Has Come," Proceedings of the IEEE, vol.28 (1990): 5-14.
17. Chandler, David, "Phase Jitter Phase noise and Voltage Controlled Crystal Oscillators," Corning Corporations retrieved from, http://www.corningfrequency.com/library/phase_jitter_note.pdf.
18. Simon, Marvin K, "A Simple Evaluation of DPSK Error Probability Performance in the Presence of Bit Timing Error," Proceedings of the IEEE, vol. 42 (1994) 263-267.
19. Lo, C.M. and Lam, W. H., "Error Probability of binary Phase Shift Keying in Nakagami-m Fading Channel with Phase Noise," Proceedings of the IEEE, vol. 36 (2002) 1773-1774.

20. Piechocki, R.J., Kasparis, C., Nix, A. R., Fletcher P. N., and McGeehan, J. P., "Bootstrap Frequency Equalization for MIMO Wireless Systems," Proceedings of IEEE, vol.7 (2003) 4175-4179.

21. Lee, T., and Donnelly, K., "A 2.5V CMOS Delay Locked Loop for an 18 MBIT 500 Megabyte DRAM," Proceedings of IEEE, vol. 29 (1994) 1491-1496.

22. Retrieved July 15 from, http://www.wirelessapplications.com/wireless/services/lostFound/images/Eb_No_and_S_N_Final.pdf

23. Retrieved August 15, 2004 from, http://jove.prohosting.com/~skripty/

www.ingramcontent.com/pod-product-compliance
Lightning Source LLC
Chambersburg PA
CBHW081730170526
45167CB00009B/3764